John Hauptman
Particle Physics Experiments at High Energy Colliders

Related Titles

Talman, R.

Accelerator X-Ray Sources

2006
ISBN 978-3-527-40590-9

Martin, B.

Nuclear and Particle Physics
An Introduction

2006
ISBN 978-0-470-01999-3

Stacey, W.M.

Fusion Plasma Physics

2005
ISBN 978-3-527-40586-2

John Hauptman

Particle Physics Experiments at High Energy Colliders

WILEY-VCH Verlag GmbH & Co. KGaA

The Author

Prof. John Hauptman
Iowa State University
Physics and Astronomy
Ames, Iowa 50011-3160
USA

Cover Image

With kind thanks to Alexander Mikhailichenko for inventing the dual solenoid, and Andrea Gaddi and Hubert Gerwig, CERN for this drawing.

Library of Congress Card No.: applied for

British Library Cataloguing-in-Publication Data: A catalogue record for this book is available from the British Library.

Bibliographic information published by the Deutsche Nationalbibliothek
The Deutsche Nationalbibliothek lists this publication in the Deutsche Nationalbibliografie; detailed bibliographic data are available on the Internet at http://dnb.d-nb.de.

Typesetting le-tex publishing services GmbH, Leipzig
Printing and Binding betz-druck GmbH, Darmstadt
Cover Design Adam Design, Weinheim

Printed in the Federal Republic of Germany
Printed on acid-free paper

ISBN 978-3-527-40825-2

Contents

Particle Physics Experiments at High Energy Colliders. John Hauptman

ISBN: 978-3-527-40825-2

Preface

"It is a privilege to do high energy physics."
– Robert Leacock

My intent is to identify and describe the general principles upon which the large detectors of experimental high energy physics are designed and built and to simultaneously start from the simplest physical principles that determine the basic measurements of position, momentum, time, and energy that, in turn, develop into the somewhat complex technological realizations in a large detector. Whenever possible I will develop "rules of thumb" that all physicists use in the early conceptual thinking stages of a design.

To my knowledge a book like this has never been written, and for good reason. The information burden is huge, and the description of even a single detector often consists of a 1000-page technical design report (TDR) backed up by engineering drawings and beam tests and involving possibly conflicting memories of events and motivations for decisions. Comparing and contrasting the designs of several detectors, which was an original goal, might be impossible, at least for me. It might also be impossible to convey the intricacies and motivations within a large group of physicists for the choices, and compromises, that are made. Often a detector design will depend on anticipated future developments in electronics, instrumentation, or materials science, about which there are incomplete and even wrong understandings. There will be competition among individuals for preferred technologies or for funding or leadership within the larger group. There will be choices and preferences that depend upon scientific judgments of likely outcomes of a test or an experiment, or that are motivated by theoretical prejudices, or that are restricted by engineering constraints or costs. Not to be neglected are the conflicts of personalities and psychology, group dynamics, the character of leadership, and the technological and personnel choices that are made in order to secure stable funding for a big detector.

There will be a good deal of detector technology discussed here, building on the fundamental physics of particles of all kinds interacting in simple ways with the atoms of which detectors are built. But the main goal is a conceptual guide to the design of the large detectors that sit in the colliding beams at today's high energy physics laboratories. Usually, only a small fraction of the physicists and students

Particle Physics Experiments at High Energy Colliders. John Hauptman

ISBN: 978-3-527-40825-2

who work on a big experiment participate in the early design of detector geometry, calorimetry, and magnetic field configuration. Beyond a certain point, these major decisions cannot be undone, while the lesser choices of tracking system or vertex chamber technology can be delayed, or even changed, far into the development phase of a detector, which in current detectors is 10 to 15 years long.

Detectors are positioned directly between the theoretical expectations for the underlying physics and the actual high energy particles delivered and colliding at the big machines. They are the main connection between the expensive machine hardware (radio-frequency cavities, dipole magnets, civil infrastructure) and the Feynman diagrams of the theories. These three aspects – theoretical understandings, experimental instrumentation, and the large facilities such as the accelerators and colliders – are a healthy triplet, and every forefront field of scientific research will have all three, or some variation of these three. In a real sense, in high energy physics today it is the detectors that provide the glue that holds the field together as a major scientific enterprise.

Big detectors are our subject, but taken alone they are purposeless and uninteresting. In the USA, detector development *per se* is seldom supported except within the context of an experiment, along with a theoretical understanding or justification for the experiment. At this moment, a Muon Accelerator Program (MAP) is under study in the USA and one of its longer-range goals is the development of a Muon Collider for $\mu^+\mu^-$ collisions at multi-TeV energies (Section 5.3.3). In addition to the essential and highly challenging machine physics, a physics justification is required that includes conceivable detectors.

Each frontier machine is unique in its technology for acceleration, resulting in interaction rates and bunch structures that differ by many orders of magnitude among machines, requiring widely varying demands on detectors. The fundamental interactions to be studied in these colliders are the interactions of the quarks and leptons of the standard model, Figure 2.1, most commonly qq (in proton–proton scattering), $\bar{q}q$ (in antiproton–proton scattering), e^-q (in electron–proton scattering), and e^+e^- or e^-e^-, and it hardly matters which ones are chosen as long as the particles can be accelerated to high enough energies and detectors can be built in which the leptons, gauge bosons, and jets (from the fragmentation of quarks) can be measured. Necessarily, we will discuss the main characteristics and descriptions of particle accelerators and their delivery of beams to a detector, along with the standard model and its particles.

Accelerators and Colliders These "machines" drive elementary particles to energies far above atomic and nuclear energies, and above the known masses of the W, Z bosons and t (top) quark. The high energies allow both the pair-production of high-mass states above $E_{\mathrm{CM}} = 2Mc^2$ and also a potential theoretical simplification of the interactions in the limit where $M_z \sim M_W \sim M_t \ll E_{\mathrm{CM}}$. These accelerated particles are always stable, limiting collisions to only four particles: $p, \bar{p}, e^+$, and e^-. In p and $\bar{p}$ collisions, it is the scattering of the quarks and gluons inside them that comprises the theoretically fundamental interactions. The scattering of electrons ($e^{\pm}$) is already fundamental since the electron is one of the partons of the standard model. The scattering of electrons from protons ($e^- p$) is fundamentally electron-quark ($e^- q$) scattering. The choice of what particles to scatter and at what energy is determined by the cleverness of the accelerator builders and Maxwell's equations[1] and by available funding.

Detectors These instruments measure the energies and momenta of the high energy reaction products from which fundamental interactions are tested and inferred. The design and building of these large and complex intruments is the main topic of this book.

Theoretical understanding A theory either guides (predictions) or explains (postdictions) the results of experiments, for example, Rutherford explained the results of his α particle scattering from *Au* nuclei with a theory, and Gell-Mann explained with only three quarks (u, d, s) the zoo of strange particles found in experiments. On the other hand, the charm quark (c) and the W and Z bosons were predicted to exist on purely theoretical grounds and subsequently found in experiments. To estimate the value of theoretical understanding, imagine a world without Dirac's prediction of antimatter. Would Carl Anderson have put a Pb plate inside his cloud chamber, resulting in the observation of a "positive electron" moving upwards, opposite to the direction of cosmic rays? Would the Bevatron have been built at a proton beam energy of 6.3 GeV[2] to produce the antiproton? At this moment, the justification for building the Large Hadron Collider (LHC) has been largely based on the theoretical expectations of observing the Higgs boson, supersymmetric particles, or several other speculations.

During those many years when we are stuck in school, we mostly learn theoretical ideas and the formal solutions to many problems in electrodynamics, mechanics, modern optics, and quantum mechanics. This is all good. However, the design, understanding, and construction of a complex detector depends on much more. It will be impossible for me to provide this in a simple book, but there is no substitute for building and beam testing instruments, however small, to gain an understanding of the parts of large detectors. Even better would be students and postdocs who are calibrating and analyzing data on ATLAS and CMS to take a year off to work on CLIC problems, or Muon Collider problems.[3] In addition, there have

1) It is a small conceit of experimentalists and theorists that we "ordered up" the machine we are using, but in reality the machines we use in experiments are those that *can be built* with current understanding, technology, and funds and are limited purely by the accelerator physicists who receive a disproportionately small share of the "credit" for discoveries.

2) Leon Lederman subsequently showed that Fermi motion in the nuclear target allows antiproton production at $\sim$ 2 GeV lower beam energy.

3) Very early in the days of cyclotrons and small synchrotrons, the people who built the machines and beam lines also did the experiments. Then came a period of bigger machines and larger experiments when experimenters knew little about the machine, and, likewise the builders of the machines knew little about experimentation. The more sophisticated machines (ILC, CLIC, Muon Collider) require a tighter coupling through the so-called machine-detector interface (MDI) between the characteristics of the machine and the design of the detector.

been many hands-on instrumentation summer schools for graduate students and early career physicists.[4]

Most big experiments are started by senior physicists, usually at the big labs (CERN, KEK, Fermilab, and SLAC), and younger physicists usually do not participate, possibly for many reasons: they lack experience and contacts, they are busy with courses and experiments, and work in this area does not easily lead to publications. However, I believe this is exactly the stage where young physicists and students should start out in high energy physics, and we have done this, to the degree possible, on the 4th detector concept and in the DREAM instrumentation collaboration. Young physicists, and even students, early in their career should work on a detector concurrent with developing a necessary understanding of the collider and its environment. At the moment, three areas offer this possibility, the International Linear Collider [1], the CERN Compact Linear Collider [2], and the Muon Collider [3], and there are active detector groups at all three. In fact, it is a goal of this book to provide a simple start on this fascinating path.

High energy physics is considered to be an arcane specialty, but I believe that the techniques and strategies involving risky R&D, technical judgments, and scientific coordination are widely applicable in industry, large astrophysics experiments, and other large scientific programs. An excellent case, and also an excellent and interesting story that illustrates all of these, is COBE, the Cosmic Microwave Background Explorer, led by John C. Mather and told in his book *The Very First Light: The True Inside Story of the Scientific Journey Back to the Dawn of the Universe* [4]. A similar excellent and recent book that narrates the historical and cultural context that led up to the big high energy experiments at the LHC is by Paul Halpern, *Collider: The Search for the World's Smallest Particles* [5]. An excellent and very readable book on accelerators is *Engines of Discovery: A Century of Particle Accelerators*, by Andrew Sessler and Edmund Wilson [6].

Acknowledgements

I thank my friends and colleagues who taught me physics and, above all, its appreciation: George Trilling in physics of all kinds, and much more, whose most endearing habit in response to a physics argument that is completely wrong is to say "I don't understand that"; L. Jackson Laslett in accelerator physics on the Electron Ring Accelerator; David Nygren in TPCs and new ideas galore; John Learned in DUMAND; and Nural Akchurin and Richard Wigmans in calorimetry, which will be a dominant component of this book and in future detectors.

Alexander Mikhailichenko has been a most invaluable collaborator, having built accelerators from scratch and understanding their relation to detectors, including the first ideas of dual solenoids in detectors and their benefits to accelerators; Se-

4) The Instrumentation Panel of the International Committee for Future Accelerators (ICFA) is now seeking a venue to fill this vacuum: "the ICFA Instrumentation Panel is considering the organization of a school devoted to in-depth studies of the various aspects of detectors and instrumentation in HEP; the school would be augmented by extensive hands-on laboratory courses." (See EDIT 2011, "Excellence in Detectors Instrumentation and Technology", A. Cattai).

hwook Lee, the first student to work on 4th; Franco Grancagnolo with his deep understanding of tracking systems; G.P. Yeh, Marcel Demarteau, and Chris Damerell in the linear collider community; Giovanni Tassielli, Vito Di Benedetto, Anna Mazzacane, Corrado Gatto, Emi Cavallo, Guiseppina Terraciano, Marco Peccarisi, and Matteo Russo and their superb detector simulation and reconstruction and analysis of physics events and, physics majors Robert Holliday, who produced most of the original graphics using the PGFPLOTS package [7], along with Priscila Torres. Many new ideas emerged in the concept detector, from Andrea Gaddi, Hubert Gerwig, Richard Wigmans, Mas Wake, and Ryuji Yamada, and new ideas in particle identification and calorimetry.

Encouragement and support came from Paul Karchin, Shahla Sharifi, and Cynthia Pasant, I am indebted to Cynthia Pasant and Robert Holliday for reading the manuscript. I dedicate this book to Miriam and Michael.

Ames, Iowa, September 2010 *John Hauptman*

1
Introduction

"Now I see. The experimentalist connects the nut and the bolt to the Feynman diagram."

– Sung Keun Park (student)

Essentially, in high energy physics we are repeating the Rutherford α-particle scattering experiment over and over in which a projectile of energy far larger than the binding energies of a system is used to probe the structure of that system, in that case, the gold (Au) atom. The energies and techniques have changed, but the scientific methods have not. Rutherford used a "natural accelerator", a radioactive α-emitter, collimated so that the α particles were aimed at a thin Au foil target. Accelerator physicists have conceived and invented, designed and built, and brought into operation a most impressive array of "machines" that accelerate particles of many kinds, by many different mechanisms, and to extraordinary energies now in the multi-TeV range. These machines, often little known in the world outside particle physics although they have been used in cancer therapy since Lawrence's cyclotron, are now used extensively in materials and biological characterizations at intense "light sources" and are now proposed for use in the deactivation of radioactive waste and in nonuranium and nonplutonium nuclear power generation [8]. Accelerators are now receiving a certain degree of national and international exposure[5] and popular notice in the case of the Large Hadron Collider (LHC) that is not always positive but that may serve to illustrate that frontier technologies are neither easy nor risk free.

In addition to science and technology, accelerators have provided a platform for international collaboration and cooperation, even, or especially, between nations whose governments are mutually hostile, such as the exchange of Soviet and American physicists between Serpukhov, Dubna, and Fermilab[6] in the 1970s. The

5) For example, the "Symposium on Accelerators for America's Future", 26 October 2009 Washington, DC. A very readable report of this Symposium is available at http://www.acceleratorsamerica.org/files/Report.pdf.

6) A prime mover was Robert R. Wilson, Fermilab director, in which the E. Tsyganov group from Dubna, USSR, worked at Fermilab and the D. Drickey group from UCLA worked at Serpukhov measuring the charge radius of the π^- and K^- mesons by direct elastic scattering, $\pi^- e^- \to \pi^- e^-$ and $K^- e^- \to K^- e^-$ on the atomic electrons in a hydrogen target.

Particle Physics Experiments at High Energy Colliders. John Hauptman

ISBN: 978-3-527-40825-2

largest and most successful international platform is the CERN laboratory and the LHC experiments on which many thousands of physicists and students worldwide work together.[7)]

1.1
Collider Energy

M. Stanley Livingston was the first student of Ernest Lawrence to build the first cyclotrons, and in the early years huge leaps were made in accelerated proton laboratory energies. Livingston plotted this laboratory energy vs. calendar year, and any plot like this is still called a "Livingston plot", shown in Figure 1.1 as an adaptation of one by Panofsky [12] and includes most accelerators from the rectifier generator in 1930 to the LHC today, plus the possible future colliders ILC (International Linear Collider), CLIC (CERN Compact Linear Collider), and Muon Collider.

An alternate version of the Livingston plot is shown in Figure 1.2, where the vertical axis is the center-of-mass energy of the fundamental constituents, the electrons, and the quarks. Like its later cousin, Moore's Law in semiconductor technology for the number of transistors on a chip, the achievements in actual laboratory particle energy are exponential in time, with continual predictions over decades that this rate cannot be maintained for long. A moment's reflection on the Livingston plot teaches us that engineering an old technology always loses out to the invention of a new one. This can be seen in the progress made with rectifier generators from 1930 to 1950 of only a factor of 10 through engineering improvements, whereas the cyclotron immediately bettered the rectifier generator by a factor of ten, and picked up another factor of ten in a decade. The proton synchrotron (weak focusing with a square-meter-sized vacuum channel for the accelerator beam) started at 20 times the energy of the highest energy cyclotron, gained a factor of 10 with strong focusing, another factor of 10 with superconducting magnets, and is poised for another factor of seven in future years. A similar cascade of electron accelerators (which began with cathode ray tubes around 1900) proceeded through Wideröe's linac, the klystrons of Stanford/SLAC, and current ILC [1] acceleration cavities of 35 MV/m (superconducting, Tesla design [13]) and the current CLIC [2] cavities of 100 MV/m (normal-temperature, two-beam CERN design [14]). The advent of colliding beams in useful experiments depended upon the achievement of beams of very small transverse dimensions for high luminosity using a variety of very clever transverse momentum cooling techniques, and this allowed all the laboratory energy of both colliding particles to contribute directly to the center-of-mass energy of the collision, and therefore all high energy accelerators today are colliders of electrons or protons.

7) See [9]. For two views of the LHC, see Evans [10] and Lincoln [11].

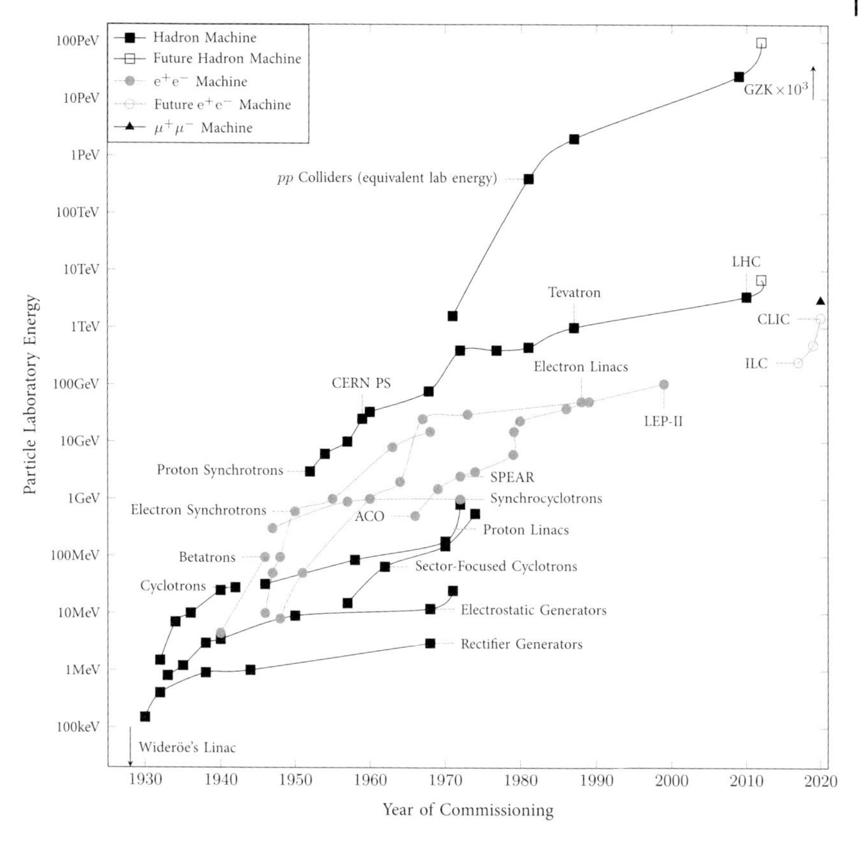

Figure 1.1 A Livingston plot from Wideröe's linac to the GZK limit. Solid symbols are for existing machines with actual particle laboratory energies on the vertical axis. Open symbols are for nonexistent machines, or for the equivalent laboratory energy of a proton that would give the same center-of-mass energy as a pp collider.

Colliders Are Near The GZK Limit

The Greisen–Zatsepin–Kuzmin (GZK) limit, 5×10^{19} eV, is the maximum energy proton that can reach Earth from far away, given that higher energy protons would be degraded by proton-γ interactions with the Cosmic Microwave Background (CMB) that fills the universe with 2.7 K photons of energy $E_\gamma \sim 2 \cdot 10^{-4}$ eV through the resonant reactions $p\gamma \rightarrow \Delta^+ \rightarrow p\pi^0$ and $p\gamma \rightarrow \Delta^+ \rightarrow n\pi^+$. The line labeled "Proton Colliders (equivalent energy)" enables a direct comparison of the laboratory energies, E_{lab}, of cosmic ray protons hitting the Earth with the man-made collisions of protons in colliders such as the LHC at center-of-mass energies $\sqrt{s} = \sqrt{2m_p E_{\text{lab}}}$. In Figure 1.1, proton colliders fall short of the GZK limit by only a factor of 500, or a factor of $\sqrt{500} \approx 22$ in collider beam energy. Given the density of the CMB photons, the mean free path of a proton at the limiting energy is 160 Mly. It won't be

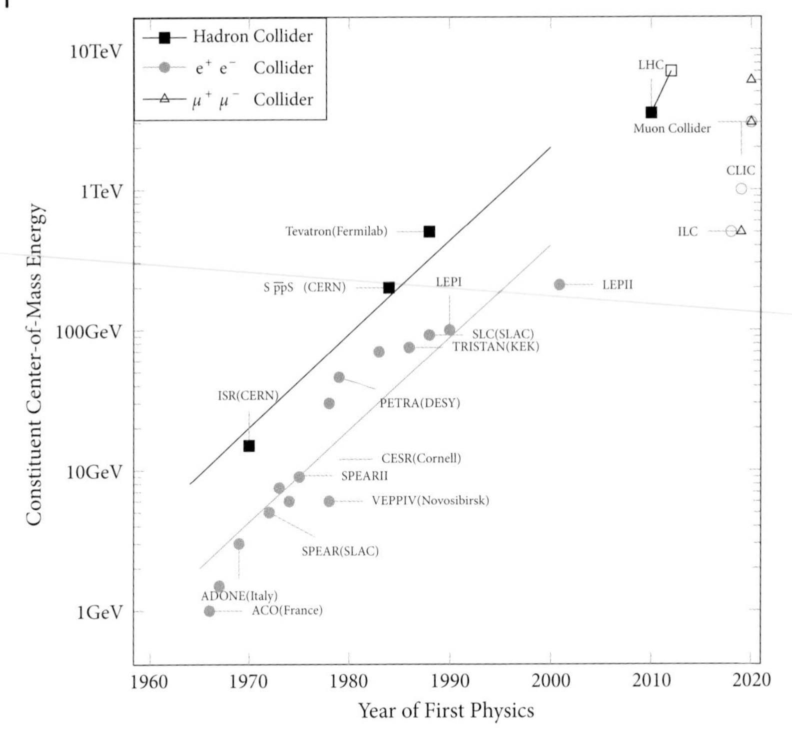

Figure 1.2 Livingston plot in constituent center-of-mass energy, $q\bar{q}$, qq, or ee.

long before humankind builds an accelerator that collides protons at center-of-mass energies higher than the highest energy cosmic rays that reach Earth, and then we won't have to listen to our cosmic ray friends brag about their accelerator that is "always on" and that beats our puny accelerators by many orders of magnitude in energy.

1.2 Collider Intensity

The "Panofsky" plot [12] of collider luminosity vs. energy in Figure 1.3 for e^+e^-, hadron, and one e^-p collider shows that achieved luminosities are roughly independent of machine energies, whereas one might expect a quadratic increase with energy shown by the dotted line due to invariant emittance conservation. The luminosity ($\mathcal{L}$) of a machine determines the interaction rate as

$$\text{Rate (interactions per second)} = \mathcal{L} \cdot \sigma_{\text{interaction}} \tag{1.1}$$

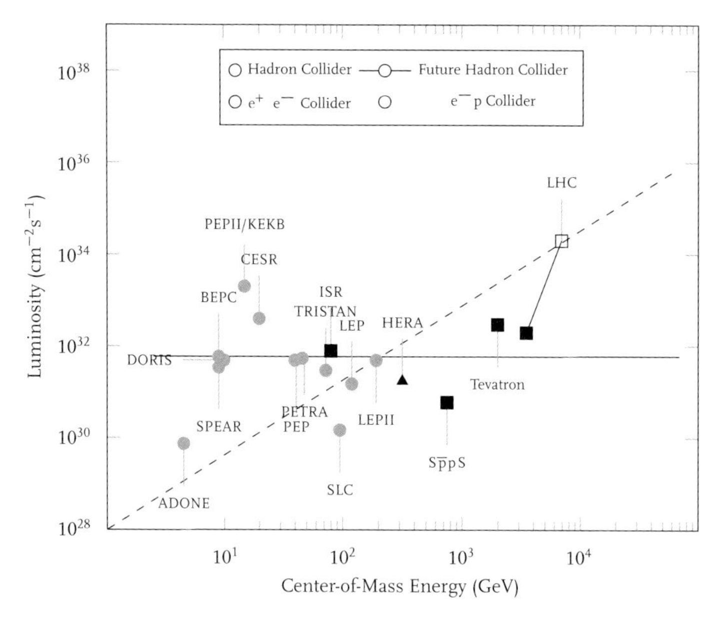

Figure 1.3 Luminosity of proton and electron colliders as a function of their center-of-mass energies.

and the luminosity is in units of $\mathcal{L}$ ($\text{cm}^{-2}\text{s}^{-1}$). If the two colliding beam bunches have n_1 and n_2 particles, the rms geometrical overlap of the bunches as they cross is described by σ_x and σ_y, and the bunches pass through each other with collision frequency f; then the luminosity is

$$\mathcal{L} = f \frac{n_1 n_2}{4\pi\sigma_x\sigma_y} . \tag{1.2}$$

The luminosity can be estimated from Figure 1.4, where the approaching particle has a total cross-section of σ_{int} with a particle in the bunch. This single particle will effectively see a fraction of the area of the opposing bunch ($A = 4\pi\sigma_x\sigma_y$) with N particles in it equal to $N \cdot \sigma_{\text{int}}/(4\pi\sigma_x\sigma_y)$. The total number of interactions for two similar bunches, each with N particles, is $N^2\sigma_{\text{int}}/(4\pi\sigma_x\sigma_y)$, and if the frequency of bunch-bunch collisions is f, then

$$\text{Rate} = f \frac{N^2}{4\pi\sigma_x\sigma_y} \sigma_{\text{int}} .$$

The luminosity is defined as the interaction rate per unit cross-section, so

$$\mathcal{L} = \frac{\text{Rate}}{\sigma_{\text{int}}} = f \frac{N^2}{4\pi\sigma_x\sigma_y} .$$

The time interval between beam crossings, $T = 1/f$, is determined largely by the type and size of the machine, for example, the CERN Compact Linear Collider

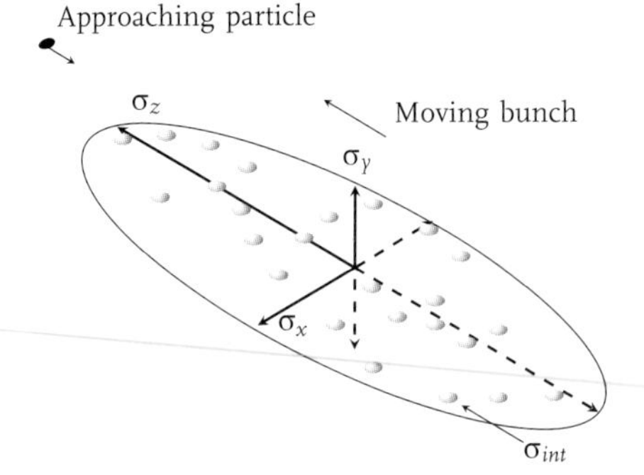

Figure 1.4 Illustration of the collision of a single particle with a "bunch" of particles. The interaction cross-section of individual particles is σ_{int}.

(CLIC) expects $1/f \approx 0.4$ ns, whereas the SLAC Linear Collider [15] (SLC) was operated at $1/f = 8.3$ ms, differing by more than a factor of 10^7. This would lead to profound differences in the choice of detectors and technologies.

Luminosity Units

The number of events produced per second in a process with cross-section σ_{event} is

$$\text{Event rate}\left(\text{s}^{-1}\right) = \sigma_{\text{event}}\left(\text{cm}^2\right) \times \mathcal{L}\left(\text{cm}^{-2}\text{s}^{-1}\right) .$$

A collider luminosity in these units, say, $\mathcal{L} = 10^{33}\text{cm}^{-2}\text{s}^{-1}$, can be written as $\mathcal{L} = 1\text{nb}^{-1}\text{s}^{-1}$ (since $1\text{b} = 10^{-28}\text{m}^2$). To illustrate, the t quark cross-section at the 14 TeV LHC is $\sigma_t \approx 1\,\text{nb}$, so at this luminosity the event rate is $\sigma_t\,(\text{nb}) \times \mathcal{L}\,(\text{nb}^{-1}\text{s}^{-1}) \approx 1\,\text{event/s}$. The "integrated luminosity" is summed over running time (T) and is $L\,(\text{nb}^{-1}) = \int_T \mathcal{L}\,(\text{nb}^{-1}\text{s}^{-1})dt$. A "standard accelerator year" is 10^7 s ($(1/\pi)^{\text{th}}$ of a calendar year), over which time the integrated luminosity becomes $L\,(\text{nb}^{-1}) \approx 10^7\text{s} \times 1\text{nb}^{-1} = 10\text{fb}^{-1}$.

1.3 Why Big Colliders and Big Detectors Are Necessary

The standard model as depicted in Figure 2.1 is so successful that the only potential discrepancies must lie either above the energies of present accelerators or below the production rates of current high intensity experiments.

Both discovery potential experiments (the highest possible beam energy) and high precision experiments (the highest possible beam intensity) are needed to test and to probe the standard model.

1.4
The Tringides Challenge

During the early Superconducting Super Collider (SSC) days, many physicists questioned the expense of the SSC (about $8.25 billion) and its purpose. A friendly condensed-matter colleague of mine, Michael Tringides, challenged me to calculate the cost of the SSC in terms of fundamental constants. It turns out this is easy to do given the standard model, the requirement that 10^4 Higgs bosons be produced in 1 year, the $\sim$ 1 eV binding energies of atoms and molecules in solids, and the cost of a kilowatt-hour of energy.

A corollary to this calculation is that the cost per foot of big colliders, either circular or linear, is the same as the cost of a condensed-matter laboratory, and this cost is about $50K /foot.[8] This is evidently so because the equipment in a condensed-matter lab or in a collider tunnel, consisting of vacuum systems, high voltage systems, monitoring equipment, computers, and so on, is basically the same. The cost of a big machine is high only because the machine is large, and it is large because achievable electric and magnetic fields are relatively small. For electric fields, $E_{\max} \approx 100$ MV/m, requiring in a linear machine an accelerating structure (half filled with $E_{\max}$) of length $L \sim 1\,\text{TeV}/(0.1\text{TeV/km}) \times 2 \sim 20\,\text{km}$. For magnetic fields, $B_{\max} \approx 10$ T, requiring a turning radius in a circular machine (half filled with $B_{\max}$) of $R = p(\text{GeV/c})/0.3B \sim (10\,\text{TeV/c})/(0.3 \times 10\,\text{T} \times 2) \sim 6$ km.

The Particle Data Group's biannual publication is a valuable resource that will be cited many times, including plots and figures that are all available at http://pdg.gov. It covers most of physics as we know it.

Radiation damage concerns are seldom treated except in the case of a specific machine, such as the LHC [16], a proton machine with a large hadronic radiation dose while running. Few books [17] address this need, while there are experts in the field, such as Mokhov [18] and Sanami [19], who can calculate these backgrounds with confidence.

I will deliberately not invoke theoretical ideas or speculations to guide the design of a detector, which is often requested by program advisory committees (PACs). In the first place, any specific physical process (Higgs production, SUSY particles with specific masses, etc.) combined with a specific *figure of merit* (mass resolution, cross-section measurement precision, ensemble efficiencies, etc.) will drive the design of a detector to a specific configuration. This configuration may have a certain magnetic field and a tracking chamber with a certain radius and a certain point precision on N points. However, a different process will drive the detector configuration to a different point, and there are many possible *figures of merit.*

This optimization process is never completed, although experimenters claim it is. It is far too arduous a task to complete, with a simulation faithful enough to ensure a correct optimization point, and, with all the possible processes and *figures of merit,* there is no unique optimum. It is essential to design the best possible detector systems for all the important particles expected in an experi-

8) Or 100K €/m, or 500K ¥/m, in 2010 financial units.

ment (for a big collider, this means everything), to configure the whole detector so as to optimize the coordination of these detector systems, and to compromise the least where conflicts arise between detectors, for example, where to put the boundary between the tracker and the calorimeter. Having measured all particles as accurately as is technically possible, the best physics will result regardless of whether we are dealing with Higgs decays, SUSY particles, technicolor, or anything else. Of course, it is important to recognize, as the ILC community has, that $W^{\pm} \rightarrow q\bar{q}$ and $Z^0 \rightarrow q\bar{q}$ decays are critical to all anticipated physics and must be measured and reconstructed with unprecedented precision in energy and angle. There are other critical needs such as precision in impact parameter measurements for tagging the decays of mesons containing charm and bottom quarks or the need to measure muons from the decay $Z^0 \rightarrow \mu^+\mu^-$ to high precision so that the Z^0 becomes a useful four-vector against which a Higgs (through any of its decay modes) can be found in a missing mass distribution against this Z^0.

In the end, a detailed simulation of the whole detector is used to check that the expected physics goals can be achieved. It is best if one detector can do it all, but at some of the most successful e^+e^- colliders (SPEAR, PEP, LEP) and pp colliders (ISR, $Sp\bar{p}S$, TEVATRON LHC) the laboratory directors have approved a collection of *complementary* detectors with differing strengths, sometimes highly unique strengths, but also with measurement overlap for checks and redundancy. In this way the top quark was found at the Fermilab Tevatron by two very different experiments, at about the same time and with similar precisions.

1.5 Useful Units, and One-Dimensional Lorentz Transformation

Whenever possible, the fine structure constant $\alpha = e^2/(4\pi\epsilon_0\hbar c) \approx 1/137$ and the convenient product $\hbar c \approx 197\,\text{eV}\cdot\text{nm} \approx 0.2\,\text{GeV}\cdot\text{F}$ will be used to simplify formulas. The Lorentz transformation of a decay particle from the center of mass of the parent to the detector frame is illustrated in Figure 1.5, and it can be understood geometrically from the one-dimensional Lorentz transformation. For a particle with mass M and momentum $\boldsymbol{P}$ in the detector, the Lorentz transformation parameters are

$$\gamma = \frac{E}{M} \text{ and } \beta = \frac{P}{E}\,.$$

In the center-of-mass frame, the mass M decays to two less massive particles $M \rightarrow m_1 + m_2$ with m_1 at an angle θ^* with respect to the momentum $\boldsymbol{P}$. The center-of-mass momentum p^* is the same for both m_1 and m_2 and given by

$$p^* = \sqrt{\left(M^2 - (m_1 + m_2)^2\right)\left(M^2 - (m_1 - m_2)^2\right)}/2M \approx \frac{M}{2}\,.$$

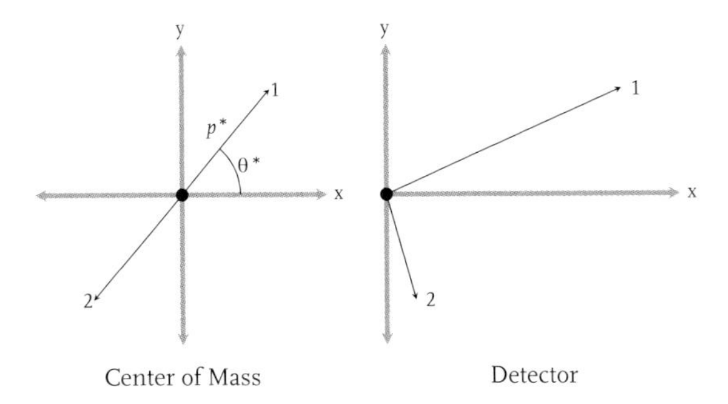

Figure 1.5 Lorentz transformation of a decay particle from the center of mass of the parent to the detector frame.

The transverse component p_{T} in the decay is invariant in a Lorentz transformation, and the longitudinal component p_{L} is boosted approximately by γ:

$$p_{\mathrm{T}} = p_{\mathrm{T}}^* \text{ and } p_{\mathrm{L}} = \gamma p_{\mathrm{L}}^* + \gamma \beta E^* .$$

For each particle m_1 and m_2, the transformations from the center of mass $(p_{\mathrm{T}}^*, p_{\mathrm{L}}^*)$ to the detector $(p_{\mathrm{T}}, p_{\mathrm{L}})$ are

$$p_{T_1} = p^* \sin\theta^* \text{ and } p_{L_1} = \gamma p^* \cos\theta^* + \gamma\beta\sqrt{p^{*2} + m_1^2}$$

and

$$p_{T_2} = -p^* \sin\theta * \text{ and } p_{L_2} = -\gamma p^* \cos\theta^* + \gamma\beta\sqrt{p^{*2} + m_2^2} .$$

Therefore, a decay particle will be boosted to higher momentum in the detector if $\cos\theta^* > 0$ or its mass m is large (making E^* large) or both. This is exactly the case in $\Lambda^0 \to p\pi^-$ decay, where p^* is small and the proton mass is large, $m_p \gg p^*$, so large, in fact, that the $\gamma\beta E^*$ term is always larger than the $p^* \cos\theta^*$ above $p_\Lambda = 0.3$ GeV/c, even for $\cos\theta^* = -1$.

$P = 0$ When M is at rest and for $M \gg m_1, m_2$, the decay products are back-to-back and $p_{\text{detector}} = p^* = M/2$ for both.

$P \gg M$ For example, for a very energetic $Z^0 \to \mu^+\mu^-$ decay (or a $\tau \to \rho\nu$ decay), where $\beta \approx 1$, the sum of the detector longitudinal momenta is

$$p_{\mathrm{L}1} + p_{\mathrm{L}2} \approx 2 \cdot \gamma\beta \cdot p^* \approx 2 \cdot \frac{E_Z}{M_Z} \cdot \frac{M_Z}{2} \approx E_Z ,$$

and the opening angle between the μ^+ and μ^- is $\tan\theta \approx p_{\mathrm{T}}/p_{\mathrm{L}}$, or

$$\theta_{\text{lab}} \approx \frac{p^*}{\gamma p^*} \approx \frac{1}{\gamma} \approx \frac{M_Z}{E_Z} .$$

This is kinematically the same as the $K^0 \to \pi^+\pi^-$ and $\pi^0 \to \gamma\gamma$ decays.

1.6 Problems

1. Estimate the proton threshold energy for the reaction $p\gamma \rightarrow \Delta^+ \rightarrow p\pi^0$, where the Δ^+ mass is 1.232 GeV/c^2 and the photon has energy equivalent to 2.7 K. At this energy, any protons in the interstellar medium are "degraded" in energy by this reaction, which, by the way, generates two very high energy γ-rays. This is the GZK limit, above which energy the universe becomes opaque to protons.

2. What should the beam energy of the Very Large Hadron Collider (VLHC; Figure 5.4) be in order to reach the GZK limit of $E_p \sim 5 \times 10^{19}$ eV.

3. For pp collisions in the LHC at 7 + 7 TeV, the pp center-of-mass energy is $\sqrt{s} = 14$ TeV. For a "fixed target" experiment in which a proton strikes a proton at rest, what must the laboratory energy of the proton be to reach the same center-of-mass energy, $\sqrt{s} = 14$ TeV? Compare to the GZK limit, and check if Figure 1.1 is correct.

4. By measuring the sagitta of the VLHC in Figure 5.4, and using as a scale the Fermilab collider with diameter 2 km, what would the center-of-mass energy of the VLHC be?

2
Particles of the "Standard Model"

"I have been asked to review the history of the formation of the Standard Model. It is natural to tell this story as a sequence of brilliant ideas and experiments, but here I will also talk about some of the misunderstandings and false starts that went along with this progress, and why some steps were not taken until long after they became possible. The study of what was not understood by scientists, or was understood wrongly, seems to me often the most interesting part of the history of science. Anyway, it is an aspect of the Standard Model with which I am very familiar, for as you will see in this talk, I shared in many of these misunderstandings."

– Steven Weinberg, "The Making of the Standard Model", CERN, 16 September 2003

Any attempt to design a big detector at a high energy collider begins with the fundamental particles of the standard model, Figure 2.1, and their available interactions. The spin-$\frac{1}{2}\hbar$ fermions are the quarks (d, u, s, c, b, t), the charged leptons (e, μ, τ), and the neutral leptons $(\nu_e, \nu_\mu, \nu_\tau)$, which are ordered from top to bottom by their electric charges. The spin-$1\hbar$ gauge bosons are the exchange particles responsible for the forces among the fermions: the massive $W^\pm, Z^0$ bosons for the "weak force", the photon γ for the electromagnetic force, and the "gluon" g for the "strong color" force.[9] The gauge bosons are shown in Figure 2.1 matched horizontally to the particles that have the necessary "charge", that is, *all* fermions carry a weak charge and exchange Ws and Zs, the quarks and charged leptons carry electric charge and exchange γs, and only the quarks carry color charge and exchange gluons. Therefore, for example, an up quark (u) can exchange any of the gauge bosons, $W^\pm, Z^0, g, \gamma$; however, for almost all detector purposes, gluon exchange completely dominates γ exchange, which in turn completely dominates $W^\pm$ and Z^0 exchange, and a quark is a strongly interacting particle for detector purposes.

Scattering any two of these fermions from each other is a good experiment. For example, $\nu_\tau \bar{\nu}_e$ scattering would be background free with only one (visible) final state, $e^+\tau^-$, but neutrinos are difficult to focus. All colliders use the u, d quarks

9) The masses of $e, \mu, W^\pm, Z^0$ are better measured than indicated in this figure. The "bare masses" of the u, d, s quarks depend on the theoretical interpretation, and the masses of ν_e, ν_μ, ν_τ are my guesses from recent oscillation data.

Particle Physics Experiments at High Energy Colliders. John Hauptman

ISBN: 978-3-527-40825-2

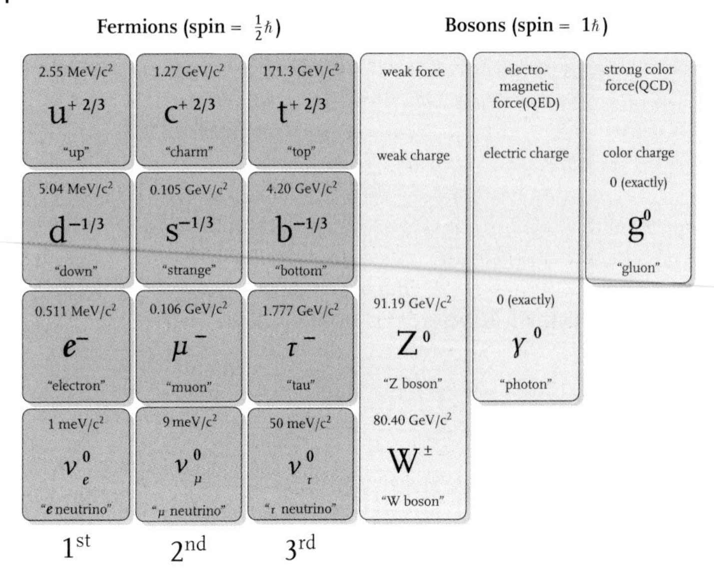

Figure 2.1 This single figure is a triumph of the human intellect. In a brief 100 years, all of these fundamental particles, and all of the interactions between them, have been discovered, studied, and understood. The discoveries have been sometimes accidental (μ^-, s, c) and sometimes deliberate (b, W, Z, t). The neutrino masses are guesses from oscillation data assuming $m_{\nu_e} < m_{\nu_\mu} < m_{\nu_\tau}$. The gauge bosons are aligned with the particles with which they couple. If the Higgs were found, where would it be placed in this diagram?

inside protons or e^+ and e^- directly in scattering all combinations of these, including antiquarks $\bar{u}$, $\bar{d}$ inside antiprotons. Only now is it deemed possible to accelerate and collide [20] μ^+ and μ^-.

We are not concerned with the fact that these particles come in three generations[10] but only how these particles interact with matter (i.e., with atoms) and how they decay into other particles.

There are hundreds of composite particles, strongly interacting hadrons, that are bound states of strongly interacting quarks, namely, mesons composed of quark-antiquark ($q\bar{q}$) pairs and baryons composed of three quarks (qqq), where q can be u, d, s, c, or b (but not t, whose large mass yields a lifetime too short for meson or baryon formation). These $q\bar{q}$ and qqq systems have excited states and are well understood in the "quark model" described, for example, in Perkins [21]. All of these

10) The three generations, or families, of quarks and leptons are a mystery: the first generation of u, d, e, ν_e is responsible for all we see and touch every day from the smallest to the largest objects. The second generation of c, s, μ, ν_μ is unnecessary but decays down into the first generation, and the third generation of t, b, τ, ν_τ is again unnecessary and decays down to the second generation. The sum of the charges of each generation is zero, a reminder that the quarks have three color charges, so the sum of the electric charges in one family is (3 colors)$\cdot[+\frac{2}{3}(u) - \frac{1}{3}(d)] - 1(e) + 0(\nu) = 0$).

states decay down into hadrons composed only of quarks from the first generation, u, d, and the strange quark s of the second generation, specifically, the proton (p) and neutron (n), the three charge states of the π meson ($\pi^{\pm}, \pi^0$), the charged and neutral K mesons ($K^{\pm}$, K^0, $\overline{K^0}$) of the meson octet, and the several baryons $\Lambda^0, \Sigma^{\pm}, \Sigma^0, \Xi^0, \Xi^-$, and Ω^- of the baryon decuplet.

For a detector, we distinguish between particles that are measured directly by the detector and those that are inferred, or reconstructed, from measured decay particles, and this distinction can be the travel distance of a particle through the detector above about a few millimeters.[11] Table 2.1 contains a list of all the particles that a detector must be able to measure ordered by mean decay length, $c\tau$.

In Table 2.1, a particle of mass m and energy-momentum E-p has Lorentz parameters $\gamma = E/m$ and $\beta = p/E$, and the laboratory mean decay length is $\gamma\beta c\tau$. The numerous resonant states (ρ, K^*, Δ, N^*, etc.) all decay down to πs, Ks, and nucleons and are left out. The D_s and B_s mesons have lifetimes similar to D and B. The $\Sigma^0 \to \Lambda^0\pi^0$ has $c\tau \sim 0.02$ nm and is left out. Clearly, the particles in the Fermi and nanometer range are only measured from their decay products; those in the hundred-μm range are found with high-precision vertex chambers; and, the millimeter- and meter-range particles are easily found with large-volume tracking chambers and calorimeters on the meter scale.

$e^{\pm}, \mu^{\pm}, \gamma$ These are the only standard-model stable particles that travel through a detector a distance of more than a few millimeters and are therefore directly spatially measurable in a tracking chamber; they are the charged leptons (e, μ) and the photon (γ). The photon is only measured after it interacts, predominantly in the electric field of a nucleus, producing an e^+e^- pair that is measured in a tracking system, or else through total absorption in a calorimeter.

ν_e, ν_μ, ν_τ Neutrinos (ν_e, ν_μ, ν_τ) travel through the detector but only interact exceedingly weakly with atoms, and for all detector purposes any neutrino is a missing three-momentum vector with zero mass.

u, d, s, g The so-called light quarks ($q = u, d, s$) and the gluon (g)[12] travel only a few Fermi (1 F $= 1$ fm $= 10^{-15}$ m $\approx$ one-proton diameter) before the color field energy density exceeds the rest masses of the numerous meson ($q\bar{q}$) states and baryon (qqq) states. At this point the color field materializes into a collimated "jet" of hadrons in a process usually called "fragmentation". These colloquial terms essentially mask our ignorance of the details by which a fundamental point-like particle turns into a collection of complicated and nonfundamental composite particles of many kinds. This simple description is akin to the "Lund" model [22],

11) The exception is the direct measurement of the $\pi^0 \to \gamma\gamma$ lifetime, where $c\tau \approx 25.1$ nm, by Jim Cronin, G. von Dardel *et al.*

12) There are eight gluons in the standard theory of quantum chromodynamics (QCD) of three color charges (*red, green, blue* with $3 \cdot 3 - 1 = 8$ independent color-anticolor states) but, apart from color counting and their couplings, which are essential in theoretical calculations of cross sections and decay rates, gluons are indistinguishable in a detector.

Table 2.1 All particles of interest for a detector, ordered by decay length, $c\tau$, in convenient units.

Particle	Fundamental (sm) or Composite (of quarks)	$c\tau$	Detector
$W^{\pm} \rightarrow$ all $q\bar{q}$ and $\ell\bar{\ell}$	sm	0.1 F	
$Z^0 \rightarrow$ all $q\bar{q}$ and $\ell\bar{\ell}$	sm	0.1 F	
$t \rightarrow W^+ b$	sm	1.0 F	
$g \rightarrow u, d, s$	sm	10 F	measure only
$u, d(q) \rightarrow \pi$s	sm (*quark*)	10 F	decay products
$s \rightarrow K$s	sm (*quark*)	10 F	
$c \rightarrow D$s	sm (*quark*)	10 F	
$b \rightarrow B$s	sm (*quark*)	10 F	
$\pi^0 \rightarrow \gamma\gamma$	composite (u, d)	25 nm	measure only decay γs
$\tau^{\pm}$	sm (*lepton*)	87 μm	
D^0	composite (c)	124 μm	
$D^{\pm}$	composite (c)	315 μm	"vertex"
B^0	composite (b)	464 μm	detector
$B^{\pm}$	composite (b)	496 μm	
$K_S^0 \rightarrow \pi\pi$	composite (s)	27 mm	
$\Lambda \rightarrow N\pi$	composite (s)	79 mm	tracking
$\Sigma^{\pm}, \Xi^{0,-}.\Omega^{-}$	composite (s)	24–87 mm	detector
$K^{\pm}$	composite (s)	3.7 m	
$\pi^{\pm}$	composite(u, d)	7.8 m	
K_L^0	composite (s)	15.5 m	
$\mu^{\pm}$	sm (*lepton*)	10^6 m	tracking and
n	composite (u, d)	10^{11} m	calorimetry
p	composite (u, d)	∞	
$e^{\pm}$	sm (ℓ)	∞	
$\nu_e, \nu_{\mu}, \nu_{\tau}$	sm (*lepton*)	∞	
γ	sm	∞	

but there are many models to describe this nonperturbative QCD process of the transition from a color-charged q or g into a jet of hadrons.[13]

These hadrons in jets are mostly pions ($\pi^{\pm}, \pi^0$), kaons ($K^{\pm}, K_L^0, K_S^0$), and nucleons (p, n) that are either produced directly in the color field or come from the decays of more massive ($q\bar{q}$) resonant states such as the vector meson nonet including ($\rho^{\pm}, \rho^0, \eta^0$) and ($K^{*\pm}, K^{*0}$), and the baryon decuplet ($\Delta^{++,+,0,-}, \Sigma^{\pm,0}, \ldots$).[14]

13) The first measurement, or discovery, of these jets was by Gail Hanson in $e^+e^- \rightarrow q\bar{q}$ events in the magnetic detector (*aka* Mark I) at the SPEAR collider [23]. The discovery of gluons materializing into jets is attributed to Sau Lan Wu in the TASSO detector at the PETRA collider [24].

14) This is the language of the static quark model; see [21].

The details of these states and their decays are not important for a detector since all of the varied decay products in the laboratory end up as photons (predominantly from $\pi^0 \to \gamma\gamma$ and $\eta^0 \to \gamma\gamma$ decays), charged tracks ($\pi^\pm$, $K^\pm$, p) from decays of resonances, and neutral hadrons (n, K^0_L, K^0_S) from $\Delta, \Sigma, \Lambda, K^*$. All of these particles are measured in the tracking and calorimeter systems of a detector and, *grosso modo*, the summed momenta of these particles is approximately the momentum[15] of the original quark (q) or gluon (g). In this way, the gluon and the three light quarks (u, d, s) are measured in common as "jets" in the detector and are, essentially, indistinguishable from each other.

c*, *b The more massive charm c and bottom b quarks fragment into mesons that contain these quarks, such as $c \to D$, where the D meson is a $(c\bar{q})$ state, and $b \to B$, where the B meson is a $(b\bar{q})$ state. There are necessarily two charge states for each D or B meson since the light quark q can be either u or d. Several excited states D^* and B^* decay rapidly (by the strong and electromagnetic interactions) down to the lowest lying D or B states, and these states are of such low mass that further decay into hadronic particles is forbidden by energy conservation. Consequently, the final decays of the D and B mesons are through weak interactions with mean lifetimes on the order of $\tau_{\text{weak}} \sim 10^{-12}$ s, or $c\tau_{\text{weak}} \approx 300\,\mu$m.

This weak mean lifetime is a very interesting number since the mean decay length in a detector is the time-dilated lifetime times the particle velocity. For a D meson of mass $m_D \sim 2\,\text{GeV/c}^2$ and a momentum ten times its mass, $p_D \sim 10 \times m_D \sim 20\,\text{GeV/c}$, the Lorentz boost factors[16] are $\gamma = E/m$ and velocity $\beta = p/E$. The mean decay length in the detector is

$$\lambda_{\text{decay length}} = (\gamma\tau_{\text{weak}}) \cdot (\beta c) = \left(\frac{E}{m_D}\tau\right) \cdot \left(\frac{p}{E}\right) = \frac{p}{m_D}\tau_{\text{weak}} \sim 0.3\,\text{mm}\,,$$

which is an easily measurable distance with a high spatial precision tracking chamber. Since these decays are very near the interaction point, or the "vertex point" of all the tracks from the interaction, these high spatial precision chambers are called "vertex chambers", and they essentially measure the impact parameters of tracks from D and B decays, and also $\tau^\pm$ leptons. There are two useful facts here: first, B mesons often decay into D mesons, and the D mesons often decay into K mesons containing a strange (s) quark. The K mesons are experimentally distinguishable from the more numerous $\pi^\pm$ mesons through either the neutral decay

15) There are profound questions about this correspondence that do not concern us, such as the fact that the quark and gluon are connected to other color-charged particles during the interaction process and are not, in fact, free particles with a momentum of their own. A more practical question concerns the fact that the measured particles of a jet have a mass (M_j) given directly by $M_j^2 = E_j^2 - (\boldsymbol{p}_j \cdot \boldsymbol{p}_j)$, where E_j is the scalar sum of the particle energies and $\boldsymbol{p}_j$ is the three-vector sum of the particle momenta of a jet. A 100-GeV gluon can fragment into a jet with mass $M_j \sim 20\,\text{GeV/c}^2$, whereas the gluon has exactly zero mass. The question is how to treat this "experimental" mass when these measured jets are themselves combined into multijet states.

16) We are using common units with $c = 1$ so that energies are measured in GeV, momenta in GeV/c, masses in GeV/c^2, and velocity in units of c, $\beta = v/c$.

$K^0 \to \pi^+\pi^-$ or by identification of the charged $K^\pm$ by its specific ionization.[17] The second useful fact is that the impact parameters of the decay products of a relativistic particle are independent of the momentum of the particle (see sidebar).

$\tau^\pm$ The $\tau^\pm$ lepton is characterized by a lifetime of $c\tau \sim 87$ μm and specific topological decays, specifically, to one-prong and three-prong final states of $e^\pm, \mu^\pm$, and $\pi^\pm$, $K^\pm$ (at 2% level), plus one neutrino (ν_τ) (or two for e, μ in the final state). Therefore, the $\tau^\pm$ have many handles of identification including the kinematics of its low mass, $m_\tau \approx 1.8$ GeV/c^2, compared to momenta at high energy colliders, so that, for example, three tracks in isolation with invariant mass below 2 GeV/c^2 is likely to be a $\tau^\pm$ decay.

Impact Parameter Is Independent of Momentum, and Approximately Equal to $c\tau$

A particle of mass m and laboratory momentum p_{LAB} has Lorentz boost factors $\gamma = E/m$ and $\beta = p/E$, $E = \sqrt{p_{\text{lab}}^2 + m^2}$, so the mean decay length ℓ for a mean lifetime τ is

$$\ell = (\gamma\tau) \cdot (\beta c) \approx \frac{p_{\text{lab}}}{m} \cdot c\tau \,.$$

For a roughly symmetric decay in the center-of-mass frame ($\theta^* \sim \pi/2$), the laboratory momentum components are $p_T \approx p_T^*$ and $p_L \approx \gamma p_L^* + \gamma\beta E^* \approx \gamma E^* \sim p_{\text{lab}}/2$, for $p_{\text{lab}} \gg m$. The laboratory angle is $\theta_{\text{lab}} \approx p_T/p_{\text{lab}}$, and the impact parameter is

$$b_{\text{impact}} \sim \ell \cdot \theta_{\text{lab}} \sim \left[\frac{p_{\text{lab}}}{m} \cdot c\tau\right] \cdot \frac{m}{p_{\text{lab}}} \sim c\tau \,,$$

independent of p_{lab}. A vertex chamber must measure the transverse coordinate of a track to a small fraction of $c\tau$. Currently, pixels of size (20 μm)^2 positioned $\ell = 2 - 10$ cm from the interaction point can achieve $\sigma_b \sim 10 - 20$ μm in the absence of multiple scattering, which contributes a term like γ_{rms}.

$W^\pm, Z^0$ The massive W and Z bosons are very democratic and decay into all available fermion-antifermion ($f\bar{f}$) pairs with nearly equal probability. For the W^+, the available final states are

$$e^+\nu_e, \mu^+\nu_\mu, \tau^+\nu_\tau, u\bar{d}(r, g, b), \text{ and } c\bar{s}(r, g, b) \,,$$

17) This is usually called a measurement of "dE/dx", that is, the ionization energy loss in the tracking detector, which is a function of $\gamma\beta = p/m$ and therefore a function of mass once the momentum is measured. See Section 3.1.1 on specific ionization measurement.

where r, g, b denote the three possible color charge combinations ($r\bar{r}$, etc.). The decay $W^+ \not\to t\bar{b}$ is not allowed by energy conservation since $m_t > m_W$. This is a total of 9 $f\bar{f}$ final states so that the branching fraction for each is approx. $\frac{1}{9} \approx 11\%$. The measured branching ratios are $\mathcal{B} \approx 10.8\%$ for $W^\pm \to \ell\nu$, and $\mathcal{B} \approx (67.6\%)/6 \approx 11.2\%$ for the six possible $q\bar{q}$ final states in $W^\pm \to q\bar{q}$.

The Z^0 is similar: the available final states are

$$e^+e^-, \mu^+\mu^-, \tau^+\tau^-, \nu_e\bar{\nu}_e, \nu_\mu\bar{\nu}_\mu, \nu_\tau\bar{\nu}_\tau \text{ for the leptons}$$

and

$$u\bar{u}(r,g,b), d\bar{d}(r,g,b), c\bar{c}(r,g,b), s\bar{s}(r,g,b), b\bar{b}(r,g,b) \text{ for the quarks}$$

for a total of 21 $Z^0 \to f\bar{f}$ final states, and the branching fraction of each is $\mathcal{B} \approx \frac{1}{21} \approx 4.8\%$. The measured branching ratios are $\mathcal{B} = 3.4\%$ for $Z^0 \to \ell^+\ell^-$ and $\mathcal{B} = (69.6\%)/15 \approx 4.6\%$ for $Z^0 \to q\bar{q}$.

Because of the color factor of 3 for each quark, the decays of W, Z into quarks, and thence into "jets", dominate the populations of W, Z final states in a collider, and, as a direct consequence, any future detector at a high precision collider must be able to measure and reconstruct the hadronic decays of $W^\pm, Z^0$ into two "jets" through their decays into quarks,

$$W^\pm \to q\bar{q} \to jj \quad Z^0 \to q\bar{q} \to jj\,,$$

which constitute nearly 70% of all $W^\pm$ and Z^0 decays. This places stringent requirements on the hadronic calorimetry, in particular its energy resolution but also its angular resolution on the two "jets", which we will discuss further in Section 3.3.2. For the first time in high energy physics, and only 30 years after their discovery in 1982 by the UA1 collaboration at the CERN $Sp\bar{p}S$ collider, the four-vectors of the $W^\pm$ and Z^0 must be measured with experimental precisions similar to those of the other fundamental partons of the standard model, such as the photon γ, the electron $e^\pm$, and the muon $\mu^\pm$.

The decays $Z^0 \to \nu\bar{\nu}$ are strictly invisible and constitute an irreproducible physical background to all processes with putative missing particles, such as the lightest supersymmetric particle (LSP) in decays of supersymmetric (SUSY) states, or particles escaping into extra dimensions.

t The last of the six quarks, t, has a huge mass $m_t \approx 172$ GeV/c^2 and therefore decays so quickly that composite hadronic mesons that would contain the t do not have time to form, and so the decay is directly into W^+ and a b quark:

$$t \to W^+ b\,.$$

Therefore, the decay final states of the t quark are as numerous, and exactly the same, as the final states of the W^+, with the addition of a b quark. The $t\bar{t}$ event can then be kinematically reconstructed by, first, reconstructing the two $W \to$

jj decays, pairing these with two suspected (or tagged by impact parameter) b quark jets, and forming the overall invariant mass of the six-jet combination. Those $W^{\pm}$ decays with a ν are missing a three-vector so that a single $t \to W^{+}b$ is not kinematically reconstructable, but by conserving energy-momentum in the whole event, a missing ν three-vector can be reconstructed.

This completes the detector measurements of all the partons of the standard model that are required for a big detector.

Everything of importance, and almost every speculation, can be observed experimentally by reconstructing these particles of the standard model, the charged leptons, the quarks, and the gluons that make "jets", those quarks whose weak decays allow lifetime tagging, and the gauge bosons. Perhaps this is a reflection of the powerful and successful standard model that most theoretical speculations do not deviate much from these decay products.

No speculative particles have been addressed here; however, any good experiment, especially one at a new energy frontier collider, must be prepared for the unexpected and even the expected unexpected. Free magnetic monopoles have not been found, although Maxwell's equations demand that they exist. Speculations such as "odderons", "unparticles", and "dark matter" particles, in addition to long-lived SUSY particles that have been proposed, must be detectable and reconstructable by any new detector.[18] Apart from brute-force better calorimeters, vertex chambers, and tracking chambers, the only domain left largely unexplored by big detectors is *time*. In Chapter 7 I will argue that all channels of all subsystems of a future detector should be clocked out in nanosecond bins, that is, digitized at 1 GHz, or faster. Even if some odd particle with a lifetime of 1 μs or ms, and a decay length of 1 mm, cm, or m were to be produced, it could be found, if not in the current event, then in the data buffers long after the primary event itself. In the 4th detector designed for the ILC with a 330-ns beam crossing time, we proposed a continuous clocked readout of all channels, from crossing to crossing. There are problems with this, not to mention the data load even after up-front zero suppression, but they are not insurmountable, and cost-effective fast digitizers will be available in the future.

2.1 Some Bubble Chamber Photographs

Bubble chamber photographs contain a wealth of visual information on the behavior of low energy particles in matter, usually liquid hydrogen (LH_2 density $\rho \approx 0.07\ \mathrm{g/cm^3}$), including visual Lorentz transformations of the decay products of

18) It is my suspicion that the next collider after the LHC will be the "last collider" and that the detectors chosen for that collider will be our "last chance" to get it all right. Of course, this notion has always proven to be wrong in the past.

weakly interacting particles on the centimeter scale.[19] Long after I finished a bubble chamber experiment with 150 000 $\Lambda^0 \to p\pi^-$ decays on film, I was reminded of their beauty when a graduate student building the TPC experiment at LBL asked me what a Λ^0 decay looked like. Everyone in Carl Anderson's cloud chamber lab and everyone in a bubble chamber group knows how the momentum asymmetry between the decay particles of K_S^0 and Λ^0 depend on the center-of-mass decay angle and lab momentum. The Lorentz transformation is visual.

Also made obvious in these photographs are the fluctuations of particles as they move through atoms. Almost all the equations we use are for average behavior, the average dE/dx, or the average rms scattering angle. But it is the fluctuations that determine the detector resolutions, not the averages, and this is particularly relevant when we discuss calorimeters where there is much confusion. Before I started simulating electromagnetic calorimeters with Electron Gamma Shower (EGS),[20] I plotted by hand the spatial distributions of low energy electrons and photons in a shower. No two showers were alike; the fluctuations were huge. Robert Wilson realized this and simulated electromagnetic showers for the purpose of calorimeter design [25, 26] long before computers became available. Let me quote his paper:

> "The procedure used was a simple graphical and mechanical one. The distance into the lead was broken into intervals of one-fifth of a radiation length (about 1 mm). The electrons or photons were followed through successive intervals and their fate in passing through a given interval was decided by spinning a wheel of chance; the fate being read from one of a family of curves drawn on a cylinder. A word about the wheel of chance: The cylinder, 4 in. outside diameter by 12 in. long, is driven by a high speed motor geared down by a ratio of 20 to 1. The motor armature is heavier than the cylinder and determines where the cylinder stops. The motor was observed to stop at random and, in so far as the cylinder is concerned, its randomness is multiplied by the gear ratio."

Wilson's Monte Carlo procedure was far more accurate than any analytical average-shower solution.

One photograph I could not find is a bubble chamber with zero magnetic field in which low energy particles slow down and stop in the LH_2, displaying the increasing multiple scattering angles, the random walk near the end, and the Bragg curve where the low-β particle is at high energy loss $dE/dx \sim 1/\beta^2$ and loses almost all of its energy in the last fraction of its range.

An anonymous statistician once said "to understand statistics, it helps to be older." I interpret that to mean that after many observations in life, one gets a feel for the averages, but also a realization of the fluctuations about the average. This is what becomes obvious from watching particles in bubble chambers.

19) University groups, e.g., Roger Barlow of the Manchester High Energy Group, use bubble chamber photographs for instructional purposes, and teachers' web sites exist with animations of particles in bubble chambers, e.g., http://teachers.web.cern.ch/teachers/archiv/HST2005/bubble_chambers/BCwebsite/index.htm.

20) EGS was written at SLAC originally for radiation shielding calculations and later became the standard for electromagnetic shower simulations in the design of calorimeters.

2.1.1
Berkeley 10-in. Bubble Chamber

The interaction is $\pi^- p \to \Lambda^0 K^0$ followed by both V^0s decaying weakly through the weak interaction: $\Lambda^0 \to p\pi^-$ and $K_S^0 \to \pi^+\pi^-$. There are stray tracks from beam π^- interacting upstream of the chamber and a couple of off-momentum beam particles. Defining x, y coordinates on the photograph from $0 \to 10$, we list the features of this photograph, labeled by (x, y); in Figure 2.2.

1. (8, 5), (7, 7), and (0, 3): δ-ray electrons from energetic ionization curl counter-clockwise; therefore the magnetic field points out of page ($\boldsymbol{F} = e\boldsymbol{v} \times \boldsymbol{B}$);

Figure 2.2 Event seen in the early Berkeley 10-in. bubble chamber exposed to a 1.1 GeV/c π^- beam. From [27].

2. $(all\ x, 0)$: Six π^- beam particles enter the chamber (two of them off-momentum). The π^- beam at 1.1 GeV/c has $\gamma\beta \approx 6$ and is near minimum ionizing (Figure 3.1);
3. at (7, 2) the π^- interacts (strong color force) with a proton, $\pi^- p \to \Lambda^0 K^0$, producing a negative strangeness Λ^0 and a positive strangeness K^0;
4. (6, 3): $K^0_S \to \pi^+\pi^-$ decays by weak interaction ($s \to W^- u$);
5. (7, 6): $\Lambda^0 \to p\pi^-$ decays by weak interaction ($s \to W^- u$); the proton is Lorentz boosted forward to high momentum in the detector;
6. the momentum of the K^0 can be estimated at $p_{K^0} \approx 0.5$ GeV/c from the geometry of its decay π^+ ($p_L \approx 0$), and so $\beta \approx 0.75$ and the mean free path is $\gamma\beta c\tau_{K^0_S} \approx 3$ cm. Since the chamber field of view is 10/in. (25 cm), this K^0_S traveled about one mean decay length. Similar estimates can be made for the Λ^0;
7. the angle and cross marks are fiducials for absolute spatial calibration from the film.

With momentum-energy conservation and identification information from dE/dx, the analysis of bubble chamber events is almost completely unambiguous. It is a pure gift to experimentalists that these particles of the standard model in Figure 2.1, and also those composite particles containing the quarks of the standard model, are obvious on the centimeter scale in detectors.

2.2 Problems

1. If a Higgs boson is discovered, and it is responsible for the masses of all the particles in Figure 2.1, where should it be positioned in that figure?

2. Estimate the mean decay lengths of the π^+ (0.23 GeV) and the μ^+ (0.14 GeV) in the BEBC bubble chamber photograph (Figure 2.3). Be inspired by Enrico Fermi who once said (approximately) "I can measure a cross section with one event, an angular distribution with two events, and a polarization with three events."

3. What should be the transverse angular granularity of an LHC forward detector in order to be able to measure the decay $Z^0 \to e^+e^-$ produced forward along the beam axis at $P_{Z_z} \sim 3$ TeV. Consider the opening angle between the e^+ and e^- and the expectation that the Z^0 mass should be measured to a precision of 10%. Make your own reasonable estimate of the energy resolution required. Is this measurement possible for the decay mode $Z^0 \to \mu^+\mu^-$ in any of the LHC detectors?

4. From the geometry of the decay $K^0_S \to \pi^+\pi^-$ in Figure 2.4, estimate the magnetic field, B.

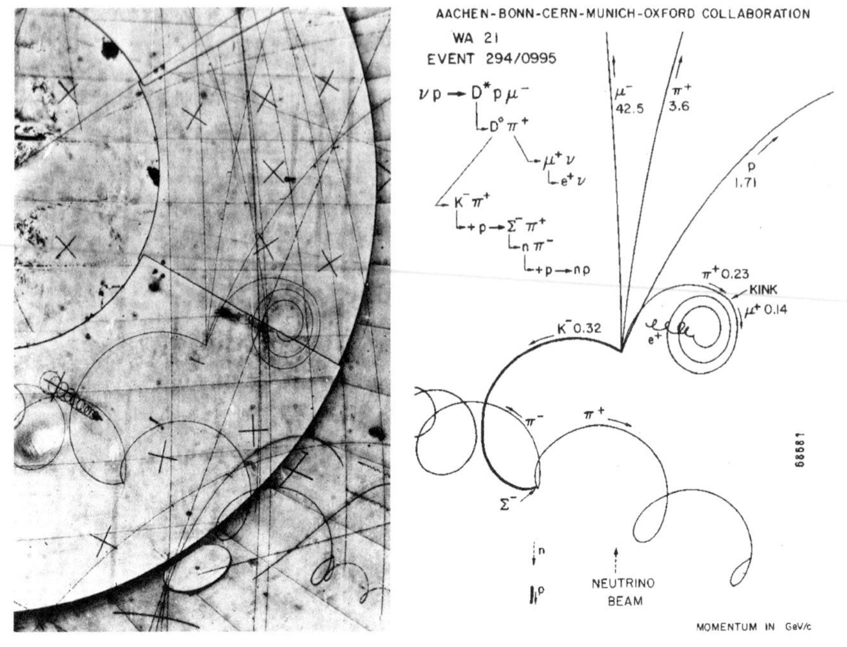

Figure 2.3 This event was photographed in the Big European Bubble Chamber (BEBC) when exposed to a ν_μ beam. The physics of this event is beautiful, and the annotations are very clear, including the "uninteresting" details such as the 0.32 GeV/c K^- from the decay chain $D^* \to D^0 \to K^-$ that stops and interacts at rest with a proton to produce a $\Sigma^-\pi^+$, the π^+ makes a helix around the $\boldsymbol{B}$ field, and the $\Sigma^- \to n\pi^-$ decays almost at rest with the π^- going vertically up and the neutron going vertically down. With luck, the neutron scatters from a proton, $np \to np$, before leaving the chamber volume. The primary physics interaction, $\nu_\mu p \to \mu^- D^{*+} p$, yields, as expected, a high momentum 42.5 GeV/c μ^- and a slow 1.71 GeV/c proton. The D^0 is boosted forward from the D^* center of mass and decays into a backward K^- and a forward π^+. The slow 0.23 GeV/c π^+ from the $D^{*+} \to D^0\pi^+$ decay travels about 1 m (in a slow 1/4-circle arc) before decaying $\pi^+ \to \mu^+\nu_\mu$, the μ^+ in turn traveling several meters (in a spiral around $\boldsymbol{B}$) before decaying $\mu^+ \to e^+\nu_\mu\nu_e$, and the very low momentum e^+ spirals along the $\boldsymbol{B}$ field lines. From this one event (and the BEBC 2-m radius) one can estimate the π^+ and μ^+ mean lifetimes.

5. The D0 measurement of a large asymmetry in like-sign di-muons (arXiv:1005.2757 [hep-ex]) suggests CP violation in the B_s and, to some, the possible existence of a fourth generation of quarks and leptons. Speculate on the features of the t' and b' quarks (their masses and decay modes) and on the two leptons (their masses and decay modes). Note that the LEP precision measurement of the number of light neutrinos, $N_\nu = 2.984 \pm 0.008$, forces the fourth-generation neutrino to be very massive, probably strictly larger than $M_Z/2$. The next step would be to outline the main features of a big collider detector that would be certain to find these particles.

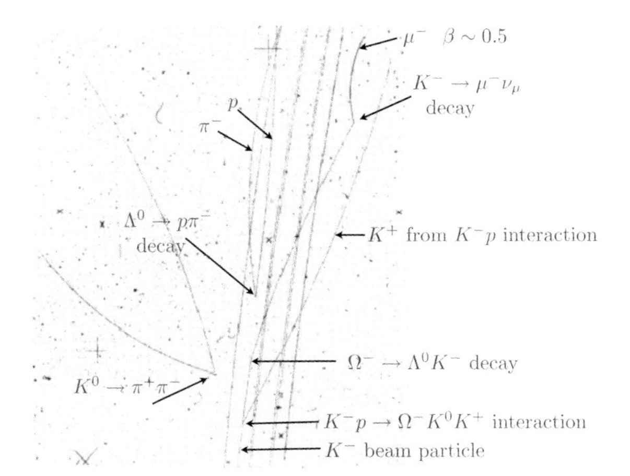

Figure 2.4 Discovery of the $\Omega^-(sss)$ three-quark state in the decay sequence $\Omega^- \to \Lambda^0 K^-$ in the BNL bubble chamber. The kinematics of this event are so clear that one event was sufficient for a discovery claim. There are four interaction or decay vertices in this event, and four equations of momentum-energy conservation at each, with no missing vectors. This one event is "overconstrained" by 16 measurements and is unambiguous.

6. Figure 2.3 literally contains two-thirds of all the particles of the standard model in Figure 2.1, plus several composite hadronic particles (both baryons and mesons), and information on lifetimes and cross-sections, multiple scattering, and energy loss, all on the spatial scale of millimeters to meters. Given that the density of liquid hydrogen is 0.07 g/cm^3 and the BEBC radius is 2 m, estimate

   a) the upper limits on the $\pi^+ p$ and $p p$ total cross-sections,
   b) the mean lifetime of the K^- and π^+,
   c) the np elastic scattering cross-section (for $p_n \sim 0.2$ GeV/c from the decay $\Sigma^- \to n\pi^-$), and
   d) an upper limit on the D^0 lifetime (by visual inspection).

3
Particle Detectors

"... the atomic hypothesis ... all things are made of atoms – little particles that move in perpetual motion, attracting each other when they are a little distance apart, but repelling upon being squeezed into each other."

– Richard Feynman, Feynman Lectures on Physics

3.1
Particles Traveling through Atoms

All detectors are made of atoms, and an understanding of detectors depends almost entirely on the electromagnetic interaction of charged particles with a tiny nucleus of large mass (Am_N) and charge $+Ze$ surrounded by a cloud of Z low-mass electrons each with charge $-e$, bound together by energies much smaller than the energies of the particles called "high energy" particles. Therefore, a particle traversing a material will experience two interactions – one with the electrons and the other with the nuclei – and both are electromagnetic. The main features of ionization energy loss ("dE/dx") from the electrons, multiple (and plural) Coulomb scattering from the nuclei, *bremsstrahlung* of γ and pair production of e^+e^- in the intense Coulomb field of a nucleus, and the enormous differences among these processes as a function of the mass of the projectile (from $e^{\pm}$ to $\mu^{\pm}$ to protons and to heavy nuclei) are all understood in this simple way. Many of the newer books on this subject gloss over the simplicities, and for this reason a long out-of-print[21] monograph of Fermi lectures is still enlightening [28]. The only important addition to this statement is that an understanding of hadronic calorimetry also depends upon an understanding of basic nuclear physics [29].

There are several excellent books on particle detectors: Leo [30] is comprehensive on detectors and techniques; Ahmed [31] is comprehensive on the physics and technical aspects of detector media; Wigmans [29] gives a thorough treatment and understanding of calorimetry; Grupen [17] covers detectors in general; Blum [32] treats tracking chambers; and Bock [33] presents a small "brief" book that is quite comprehensive, interesting, and well written. It would be foolhardy to attempt a duplication, but I will describe the basic physics from first principles in order to

21) Visit your local physics library, or just ask the oldest physicist you know.

Particle Physics Experiments at High Energy Colliders. John Hauptman

ISBN: 978-3-527-40825-2

establish the essential dependencies of momentum and energy resolutions, and spatial and time resolutions, on the basic and simplest measurements made by detectors.

3.1.1 Particle Interactions with Atomic Electrons (e^-)

Interactions with the electrons of atoms are responsible for ionization energy loss and the generation of scintillation light and Čerenkov light emission.

Ionization Energy Loss, dE/dx, Scales as $(\ln\gamma)/\beta^2$

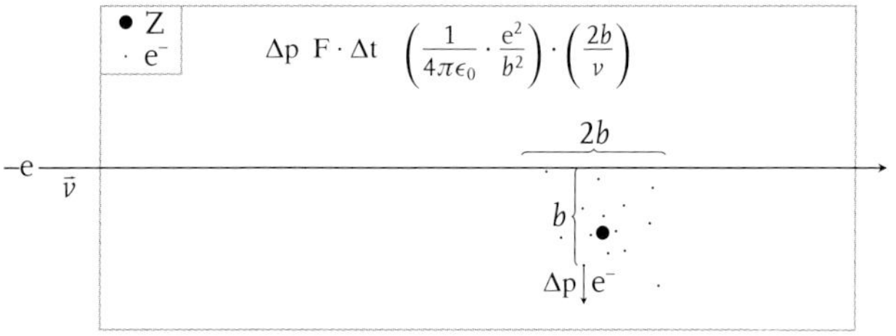

• Z · e^- $\quad \Delta p \; F\cdot\Delta t \; \left(\frac{1}{4\pi\epsilon_0}\cdot\frac{e^2}{b^2}\right)\cdot\left(\frac{2b}{v}\right)$

$2b$, $-e$, $\vec{v}$, b, Δp, e^-

A charged particle moving through atoms exerts a perpendicular Coulomb force on all charges (electrons and nuclei) characterized by the momentum impulse, or the force times the time interval,

$$\Delta p \approx \left[\frac{1}{4\pi\epsilon_0}\frac{e^2}{b^2}\right] \times \left[\frac{b}{v}\right] = \frac{\alpha\hbar c}{bv} ,$$

written simply in "natural units" with $\alpha = e^2/4\pi\epsilon_0\hbar c \approx 1/137$ and $\hbar c \approx 197\,\mathrm{eV\cdot nm}$. The massive nuclei, which receive an impulse Z times larger than the electrons, remain essentially stationary during the b/v passage period, and therefore negligible work is done on the nuclei. In contrast, the electrons are accelerated due to their small mass, and acquire a kinetic energy

$$\Delta E = \frac{(\Delta p)^2}{2m_e} = \frac{(\alpha\hbar c)^2}{2m_e c^2}\frac{1}{b^2}\frac{1}{\beta^2} ,$$

using $\beta = v/c$, which is exactly the energy lost by the passing particle by energy conservation. This displays the velocity dependence of $1/\beta^2$ seen clearly in the data of Figures 3.1a,b. The total energy loss, or dE/dx, is the sum over the number of electrons between b and $b + db$ in a length dx of matter, $\mathcal{N}_e = (Z/A)\rho \cdot 2\pi b\,db \cdot dx$, and the integral over impact parameter b yields a logarithm, $\int db/b \sim \ln(b_{\max}/b_{\min})$. The limits on b are $b_{\min} \approx \hbar/p \approx \hbar/\gamma m v$ by the Uncertainty Principle, and $b_{\max}$ is limited such that the E-field pulse frequency in the electron rest frame, $\gamma v/b$, is greater than the (mean) bound

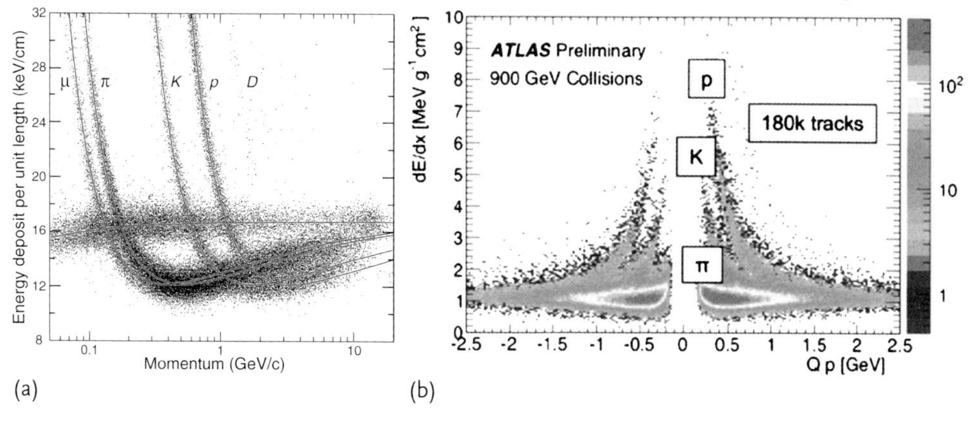

Figure 3.1 These spectacular plots show direct measurements of specific ionization energy loss for several species of particles. (a) The time projection chamber (TPC) for e, μ, π, K, p, and deuteron (D) showing explicitly the mass dependence. The deuterons in the plot were, presumably, knocked out of the Be beam pipe. The ionization energy loss is in units of keV/cm in the PEP4 TPC with a gas mixture Ar – CH_4 (90:10) at 8.5 atm pressure. The mean is calculated from a "truncated mean" that removes the δ-rays and other energetic ionization electrons from the sample, several of which are evident in the bubble chamber photograph (Figures 2.4), and (b) the dE/dx in the silicon wafers of ATLAS.

electron atomic oscillation frequency, $\bar{\nu}$, or $b_{\max} \approx \gamma v/\bar{\nu}$. For $2m_e c^2 \approx 1$ MeV,

$$dE(\text{MeV}) = \frac{(\alpha\hbar c)^2}{2m_e c^2}\frac{dx}{\beta^2}2\pi\left(\frac{Z}{A}\right)\rho\ln\left[\frac{\gamma^2\beta^2 m_e c^2}{\hbar\bar{\nu}}\right]$$

$$\approx \frac{0.15}{\beta^2}\frac{Z}{A}\ln\left[\frac{\gamma^2\beta^2 m_e c^2}{\hbar\langle\nu\rangle}\right]\cdot\rho dx\ .$$

At "minimum ionizing", as seen in Figure 3.4, $\gamma\beta \approx 3$, and for typically $\hbar\bar{\nu} \approx$ 10 eV, the logarithm is equal to about 10.8, so

$$\boxed{\frac{1}{\rho}\frac{dE}{dx} \approx 1.6\,\text{MeV/g}\cdot\text{cm}^{-2}} \quad \boxed{\text{the ionization rate for a minimum ionizing particle, or } mip} \qquad (3.1)$$

and this number ranges from 2.0 for H_2O down to 1.1 MeV/g $\cdot\, cm^{-2}$ for U. In MeV/cm, dE/dx (MeV/cm) $= \rho\,(\text{g/cm}^3)\,dE/dx$ (MeV/g $\cdot$ cm^{-2}).

Ionization energy loss, dE/dx A simple treatment of ionization energy loss is shown in the sidebar that contains all the simple essentials. An exact treatment of ionization energy loss is that by Bethe–Bloch, well described in the Particle Data

Group (PDG) [49, Section 27.2]

$$\frac{dE}{dx} = K\frac{Z}{A}\frac{1}{\beta^2}\left[\frac{1}{2}\ln\frac{2m_e c^2\beta^2\gamma^2 T_{\max}}{I^2} - \beta^2 - \frac{\delta(\beta\gamma)}{2}\right], \tag{3.2}$$

taking into account the kinematic limit on energy transfer to the low-mass electrons ($T_{\max}$), a careful treatment of the limits on the integral over impact parameters of atomic electrons (I, the ionization potential), the relativistic contraction of the transverse electric field that provides the impulse (γ^2), and the saturation of the energy loss at high γ due to the polarization of the atomic medium (δ). The last effect is seen in Figure 3.1a, where the rising energy loss for $\gamma\beta > 3$ (the logarithmic rise) saturates at high $\gamma\beta$ at about 1.4 times the minimum in this 8.5-atm Ar gas mixture. The saturation is larger in more dense media, such as silicon, as seen in Figure 3.1b, in which the logarithmic rise is not discernible. It is not obvious from the Bethe–Bloch form above, but the $\delta(\beta\gamma)$ term manages to cancel the logarithm term as $\gamma\beta \to \infty$, PDG [49, Section 27.2.2].

In a detector with finite-sized volumes in which energy losses are measured, the actual measured energy loss will depend on the details of the geometry. For example, in a gas of low density an energetic ionization electron, or δ-ray, may leave the gas volume so that the detected energy loss is less than the actual energy loss (restricted energy loss). A magnetic field will turn a low energy particle, *for example*, a shower particle in a calorimeter, into a small circle at the gyro radius $\rho = p/(0.3B)$, or drive a particle leaving an absorber back into the absorber, again reducing the measured energy loss. At very low velocities where the particle velocity is comparable to atomic electron velocities, $\beta \sim \alpha c$, the ionization rate is also reduced, and therefore a detector whose signals derive from very low energy particles will not measure a pure Bethe–Bloch energy loss. In some gases, UV photons generated in the avalanche may exit the sensitive volume before being absorbed. In scintillation media, the light yield is not proportional to the ionization energy loss in the media due to recombination of the ionization electrons with the positive ions left in the dense column of ionization. This is parameterized by Birks' law as

$$\frac{d\mathcal{L}}{dx} = \mathcal{L}_0\frac{dE}{dx}\frac{1}{(1 + k_B dE/dx)},$$

with Birks' constant typically of order $k_B \sim 0.01\,\mathrm{g \cdot MeV/cm^2}$.

For these reasons, this exact and beautiful Bethe–Bloch treatment is seldom useful in detector design. A particle ionizing atoms in a sensitive element of a calorimeter (scintillator, LAr) or a tracking layer (silicon depletion region, gas) will deposit its energy in a volume determined by many factors, including the energy of the particle. In the end, a signal taken out of a detector in millivolts (mV) or picocoulombs (pC) will be calibrated against a known energy or against the known response of a radioactive source particle.

Note that ionization energy loss is predominantly to the electrons of the atoms, while the energy loss to the nuclei is vastly smaller since $\Delta E \approx \frac{1}{2}(\Delta p)^2/(A \cdot m_N)$ for nuclei of mass number A. For material with A, Z, the ratio of ionization energy loss due to the nuclei relative to the electrons is

$$\left(\frac{dE}{dx}\right)_{\text{nuclei}} \approx \frac{Z}{A}\frac{m_e}{M_N}\left(\frac{dE}{dx}\right)_{\text{electrons}} \approx \frac{1}{4000}\left(\frac{dE}{dX}\right)_{\text{electrons}} .$$

The factor of Z comes from the original Coulomb force of the nucleus. The energy loss to nuclei is considered negligible in all detectors.

Scintillation light emission Some atomic and molecular media have energy level structures that allow excited electrons to be raised to excited states, from which they deexcite with the emission of optical photons. These atoms and molecules are excited by exactly those momentum impulses that lead to ionization, and therefore scintillation light is proportional to ionization, except for Birks' subsequent recombination. Both organic (mostly plastic) scintillators and inorganic (mostly crystal) scintillators are widely used in physics, medicine, and industry. A good description and review of their essential characteristics are contained in the PDG Review [49, Sections 28.3 and 28.4].

Since scintillation is a quantum process, and the deexcitation is an atomic transition, there is a quantum lifetime and an exponential decay time distribution to the emitted light. These lifetimes range over many orders of magnitude, but in a high energy physics experiment, speed is usually very important, and the fastest scintillators deliver their light in a fraction of a nanosecond. There are many considerations besides the lifetime such as light yield (photons/MeV), density, radiation length, wavelength emission spectrum, temperature dependence of light yield and emission spectrum, index of refraction, sensitivity to moisture, and cost.

Čerenkov light emission in a refractive optical medium Čerenkov light is generated instantaneously on the time scale of any detector by any relativistic charged particle whose velocity β exceeds the inverse of the refractive index of the medium, $\beta > 1/n$. For glasses, crystals, plastics, and water where $n \approx 1.33 \to 1.55$, Čerenkov light is ubiquitous and becomes a very interesting and important physical signal that has been used cleverly in many experiments, such as BaBar and Belle with their high requirements for particle identification. It is also used to great advantage in dual-readout calorimeters in which the Čerenkov light is mainly generated by the relativistic $e^{\pm}$ of the shower (the electromagnetic fraction), whereas the slow MeV neutrons, the nonrelativistic 100-MeV spallation protons, and even the slow $\pi^{\pm}$ and $K^{\pm}$ shower particles contribute no Čerenkov light.

Both the Čerenkov angle of photon emission and the photon intensity are functions of the particle velocity and therefore can play a powerful role in particle identification. The Čerenkov angle θ_C is given by

$$\cos\theta_C = \frac{1}{n\beta}$$

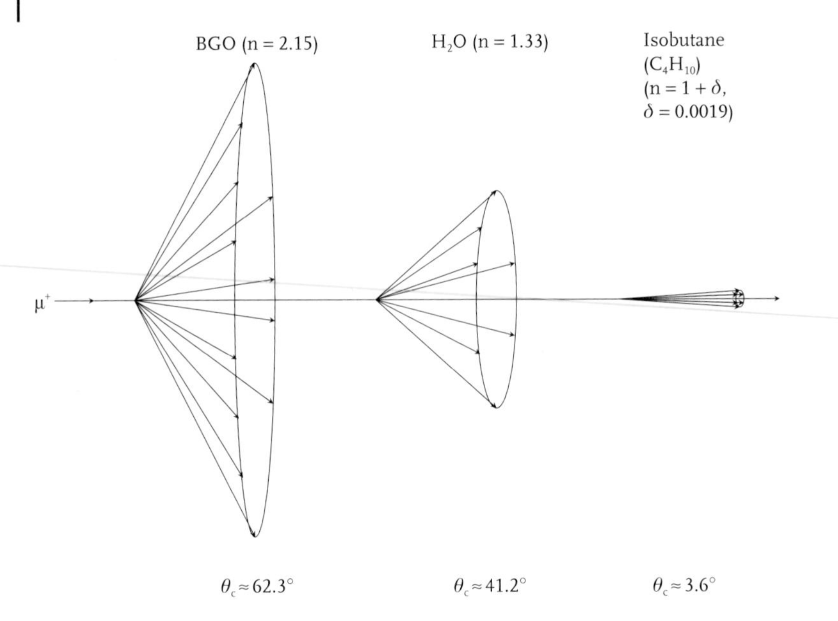

Figure 3.2 Čerenkov light as it would be generated in three optical media: a crystal of bismuth germanate (BGO, $n \approx 2.15$), water (H_2O, $n \approx 1.33$), and isobutane gas (C_4H_{10}, $n \approx 1 + \delta, \delta = 0.0019$. The Čerenkov angle and approximate intensity of photons correspond to radiators of 0.4 mm for BGO and H_2O and 40 mm for isobutane.

and the intensity in photons/(eV cm) is given by

$$\frac{d^2 N}{d E d x} = \frac{\alpha}{\hbar c} \sin^2 \theta_C \approx 370 \sin^2 \theta_C (\mathrm{eV}^{-1}\,\mathrm{cm}^{-1}) .$$

In the illustration in Figure 3.2, the Čerenkov angle and the intensity are approximately accurately depicted for these three very different media: 0.4 mm of BGO, 0.4 mm of H_2O, and 40 mm of isobutane.

For gases $n \approx 1 + \delta$ with $\delta \ll 1$, and these equations simplify since $\cos \theta_C \approx 1 - \theta_C^2/2 + \ldots$ and $1/n \approx 1 - \delta + \ldots$, so the Čerenkov angle and the photon intensity are

$$\theta_C = \sqrt{2\delta} \quad \text{and} \quad \frac{d^2 N_C}{d E d x} = \frac{\alpha}{\hbar c} 2\delta .$$

Since $\hbar c \approx 200\,\mathrm{eV} \cdot \mathrm{nm} = 2 \cdot 10^{-5}\,\mathrm{eV} \cdot \mathrm{cm}$ and $\alpha \approx 1/137$, the Čerenkov photon angle and photon flux are

$$\theta_C \approx 0.06 \text{ and } \frac{d^2 N_C}{d E d x} \sim 1\,(\text{photon/cm} \cdot \text{eV}) \text{ for } \delta \approx 0.002, \text{(isobutane)} .$$

The intensity of Čerenkov light is lower by a large factor than scintillation light, largely because of α, but nevertheless it can be an important contributor to the light signals from scintillating crystals in big detectors [34], especially if those crystals are to be calibrated to the 1% level.

The spectrum of Čerenkov light is flat in frequency, ν, and therefore goes as $1/\lambda^2$ in wavelength,

$$\frac{dN_C}{d\nu} = \text{constant or } \boxed{\frac{dN_C}{d\lambda} = \frac{1}{\lambda^2}},$$

so that most of the Čerenkov light is in the deep blue.

3.1.2
Particle Interactions with Nuclei (Ze)

Interactions with the atomic nucleus result in multiple scattering, bremsstrahlung (by electrons), pair production (of electrons), bremsstrahlung and pair production by muons, and the interactions of nuclear particles with nuclei.

Multiple Coulomb Scattering: θ_{rms} Scales As $\sqrt{\ell(X_0)}/p$

A high energy charged particle with a mass, m, much larger than the electron mass, m_e, passing through matter with nuclei of charge Ze and mass M will Coulomb scatter from the nuclei with small angular deflections given by

$$\sigma(\theta) = 4\left(\frac{Ze^2}{p\beta c}\right)^2 \frac{1}{\theta^4} = \left(\frac{2Z\alpha\hbar c}{p\beta c}\right)^2 \frac{1}{\theta^4}.$$

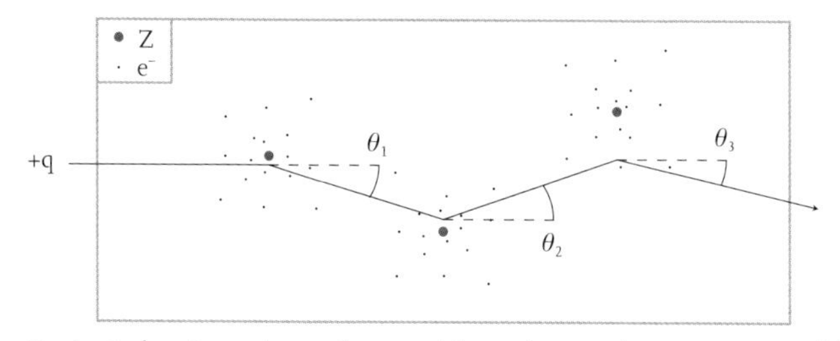

Projected onto a plane, the particle trajectory is a sequence of individual scattering angles $\theta_1, \theta_2, \theta_3, \ldots$, and so on. After a thickness of material ℓ is traversed, the total net scattered angle is $\Theta = \Sigma_i \theta_i$, and for flight paths ℓ_i beween scatters, the net transverse displacement is $Y = \Sigma_i \ell_i \theta_i$. This problem is made for the Central Limit Theorem, a remarkably powerful statistical theorem, which states that the sum of many events sampled from *any* distribution (for example, Rutherford scattering angles) will be Gaussian as long as the moments of the distribution are noninfinite, which is equivalent to saying that no part of the distribution goes off to infinity. The mean squared deflection is $\overline{\Theta^2} = \overline{\theta_1^2} + \overline{\theta_2^2} + \overline{\theta_3^2} + \cdots \approx \mathcal{N}\overline{\theta_1^2}$ since the cross-terms like $\overline{\theta_1\theta_2}$ are all zero since each scatter is independent. $\mathcal{N}$ is proportional to the depth of material, ℓ, leading to a Gaussian with rms width, θ_{rms}, that scales as $\sqrt{\ell}/p\beta$

$$\theta_{\text{rms}} \approx \frac{0.0136\ \text{GeV/c}}{p\beta}\sqrt{\frac{\ell}{X_0}}$$

or

a 1-GeV/c particle traversing 1 radiation length is typically deflected in angle by 1.4% .

Multiple scattering, θ_{rms} The exact treatment of multiple and plural Coulomb scattering is quite complicated, and the best formulation is by Molière [35][22] and described in the PDG [49, Section 27.3]. The projected scattering angle in a plane, Figure 3.3, is both positive and negative with mean zero. The mean width about zero is θ_{rms}. This root-mean-square width of this Gaussian projected angular distribution depends on the particle momentum p and the amount of matter traversed ℓ/X_0 in radiation lengths. θ_{rms} is approximately represented by

$$\theta_{\text{rms}} = \frac{0.0136\ \text{GeV}}{p\beta c}\sqrt{\frac{\ell}{X_0}}[1 + 0.038\ln(\ell/X_0)] ,$$

which includes small modifications due to plural and singular scattering effects in thin absorbers (not enough scatters to completely satisfy the Central Limit Theorem). All other quantities in Figure 3.3 are expressed in terms of θ_{rms} as

$$\begin{aligned} \psi_{\text{rms}} &= \frac{1}{\sqrt{3}}\theta_{\text{rms}} , \\ y_{\text{rms}} &= \frac{\ell}{\sqrt{3}}\theta_{\text{rms}} , \\ s_{\text{rms}} &= \frac{\ell}{4\sqrt{3}}\theta_{\text{rms}} . \end{aligned} \tag{3.3}$$

We shall see that it is the random fluctuations in the sagitta, s_{rms}, that limit the momentum resolution from the magnetic field bending sagitta of a charged track. ψ_{rms} and y_{rms} limit the impact parameter resolution of a vertex chamber. For a very thin medium, with very few Coulomb scatters, it can happen that one scatter or several scatters can be very large, often called "plural scattering", resulting in a long tail on the Gaussian.

Of course, there is multiple scattering from the electrons, but this is vastly smaller since kinematically the maximum scattering angle from an electron is $\theta_{\text{max}} \approx 1.2\, m_e/M$ and, although there are Z more individual scatters per atom, this only broadens θ_{rms} by $\sqrt{Z}$.

22) A more accessible paper is by H.A. Bethe [36]. Very recent updates verify the wide applicability of the Molière theory [37].

23) Bremsstrahlung means "braking radiation", and indeed a thin sheet of Pb dramatically brakes a high velocity electron, reducing the average electron energy to $1/e$ of its initial value in one radiation length, X_0.

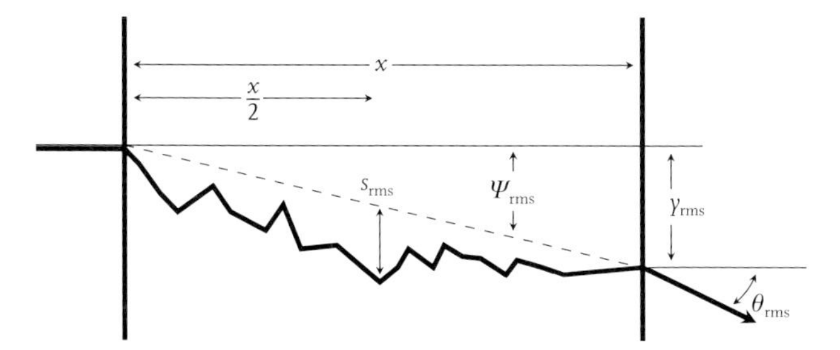

Figure 3.3 The multiple scattered path of a particle traversing ℓ/X_0 radiation lengths, with an exiting angle of θ_i, a sagitta of s_i, a transverse displacement of y_i, and a bend angle of ψ_i. The rms widths of each of these are θ_{rms}, s_{rms}, y_{rms}, and ψ_{rms}.

Bremsstrahlung by electrons from a nucleus In addition to ionization energy loss, electrons and muons also lose energy by radiation[23] when accelerated by a massive nucleus of charge Ze. This is a highly stochastic process in which an accelerated charge emits photons of energy k (cf. synchrotron radiation) in a dk/k distribution, and the mean energy of the electrons is degraded exponentially as

$$\overline{E_{e^-}} = E_0 e^{-z/X_0},$$

which serves as a definition of the radiation length, X_0, for electrons. Compared to ionization energy loss, the energy at which ionization and radiation energy losses are equal is the critical energy of the medium, which can be approximately taken to be $E_c \approx X_0 \cdot dE/dx$.

Pair production by photons from a nucleus Pair production by a photon in the Coulomb field of a nucleus is closely related to electron bremsstrahlung since in both cases a nucleus is required to supply the needed photon, which also provides the momentum transfer needed to conserve momentum and energy in the process. Since it is an electromagnetic process, the momentum transfer, Δp, is small and the energy lost to the nucleus is very small, $\Delta E = (\Delta p)^2/2M$. The mean free path for the attenuation of a photon beam due to pair production is $(9/7)X_0$.

Bremsstrahlung and Pair Production Are Responsible for Electromagnetic Showers

The development of an "electromagnetic shower" in a material is very simple: an incident electron will emit photons continuously by bremsstrahlung, $e^{\pm} \to e^{\pm}\gamma$, as it passes through atoms, occasionally emitting an energetic photon, and in one radiation length, X_0, will give up about two-thirds of its energy. Each of these photons will convert to electron-positron pairs, $\gamma \to e^+e^-$, in about one X_0. Therefore, in each X_0, there is a doubling of the number of energetic electromagnetic particles, and after n radiation lengths, the number of particles is $\sim 2^n$, an exponential increase in depth. This exponential multiplication stops when the electrons degrade to the "critical energy", E_c, where an electron is more likely to lose energy by ionization than by radiation. The photons cease pair production at the e^+e^- threshold, $E_\gamma \sim 2m_e c^2 \approx 1$ MeV. In the later stages of a shower, the photons continue to produce calorimeter signals through Compton scattering. $\gamma e^- \to \gamma e^-$, the electrons come to rest, and the positrons eventually are annihilated

Bremsstrahlung by muons The Bethe–Bloch ionization energy loss formula above is accurate to 1% for muons in Cu from 6 MeV to 6 GeV. Below this range, atomic physics suppresses energy loss when the particle velocity is comparable to the electron orbital velocity, αc. Above this range, muons and pions, and eventually every charged particle when it reaches a momentum near $\beta\gamma mc \sim 3000Mc$, begin to engage in bremsstrahlung in the high-Z Coulomb field of the nucleus. Both of these effects are shown in Figure 3.4 for the case of muons in Cu from $\beta\gamma = 0.001$ to $\beta\gamma = 10^6$, that is, muon momenta from 0.1 MeV/c to 100 TeV/c. The energy at which ionization and radiation losses are equal is called the critical energy for muons, $E_{\mu c}$, about $E_\mu \approx 300$ GeV, or about $\beta\gamma \approx 3000$.

In the CMS experiment, this curve was measured from 5 GeV to 1 TeV [38] using cosmic ray μ's at a depth of about 100 m, shown in Figure 3.5. The panel at the bottom is the ratio of the data to the simulation, as a check on the measurement. This is the same detector depicted in Figures 4.1 and 6.15.

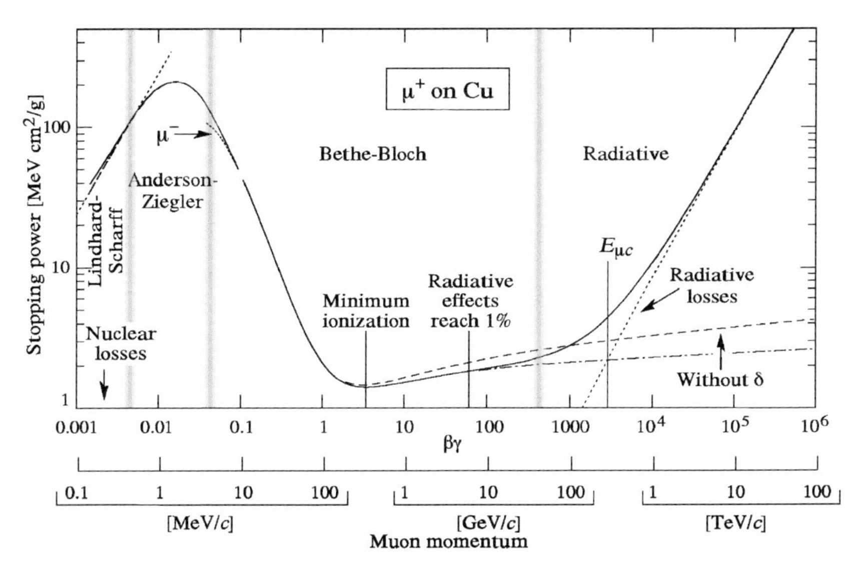

Figure 3.4 Energy loss of muons in Cu (expressed as "stopping power" in [MeV · cm²/g] from nonrelativistic to ultrarelativistic energies. From PDG [49, Figure 27.1].

More importantly, the ionization and radiation energy losses of muons have been *separately* measured for the first time in the $10\lambda_{int}$ DREAM module, consisting of a Cu absorber filled with fibers, and exposed to a μ^- beam from 40 to 300 GeV.

The dual-readout DREAM module was equipped with two kinds of fibers, scintillating fibers that sampled energy losses in 12% of the module by volume and clear fibers (both quartz and plastic) in which Čerenkov light was generated by relativistic particles, that is, predominantly $e^{\pm}$. In the quartz fibers that filled the central volume of the module and the region through which beam μ^- were directed, the numerical aperture was $NA \approx 0.33$ with a corresponding capture angle of $\theta \approx \sin^{-1} 0.33 \approx 19°$. The Čerenkov angle in the quartz fibers was $\theta_C \approx \cos^{-1}(1/1.45) \approx 46°$. In the clear plastic fibers through which the μ^- passed when the DREAM module was tilted by several degrees, the fiber capture angle was 30° and the Čerenkov angle was 48°. In both fibers and throughout the module, the Čerenkov light was generated outside the numerical aperture of the fibers and, therefore, the Čerenkov signal was zero for a nonradiating muon.

Since the scintillation signal (S) and the Čerenkov signal (C) were calibrated with a 40-GeV e^- beam, the purely electromagnetic radiative component results in $S \approx C$ and, therefore, the difference $S - C \approx dE/dx$, that is, just the ionization energy loss of the DREAM module. The scintillation response to μ^- at 40, 100, and 200 GeV is shown in Figure 3.6a, with the very characteristic dk/k radiation spectrum that is expected. A comparison of the scintillation and Čerenkov distributions are shown for 200 GeV in Figure 3.6b, in which the means are 3.46 and 2.37 GeV, respectively.

These mean energy losses for scintillation and Čerenkov are plotted against the μ^- beam energy in Figure 3.7, in which the means each increase logarithmically with muon energy as expected for a dk/k radiative distribution, but the difference $(S - C)$ is constant at about 1.1 GeV.

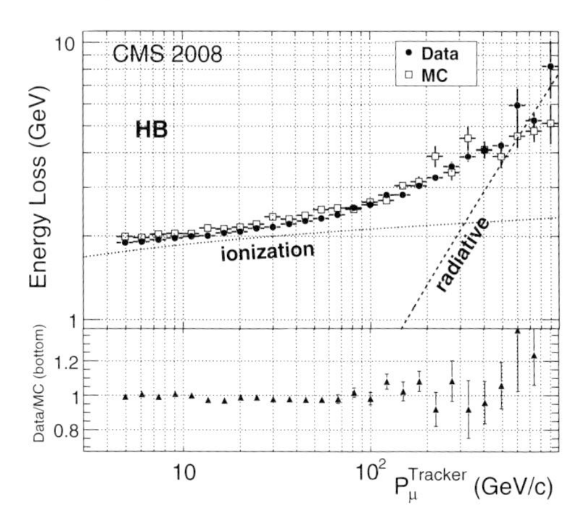

Figure 3.5 Measured muon energy loss in CMS from 5 GeV to 1 TeV [38].

This separation of the calorimeter responses into purely radiative (C signal in Figure 3.6b lower plot), and radiation-plus-ionization (S signals in Figure 3.6a and b upper plot) is the first time this has been directly measured.

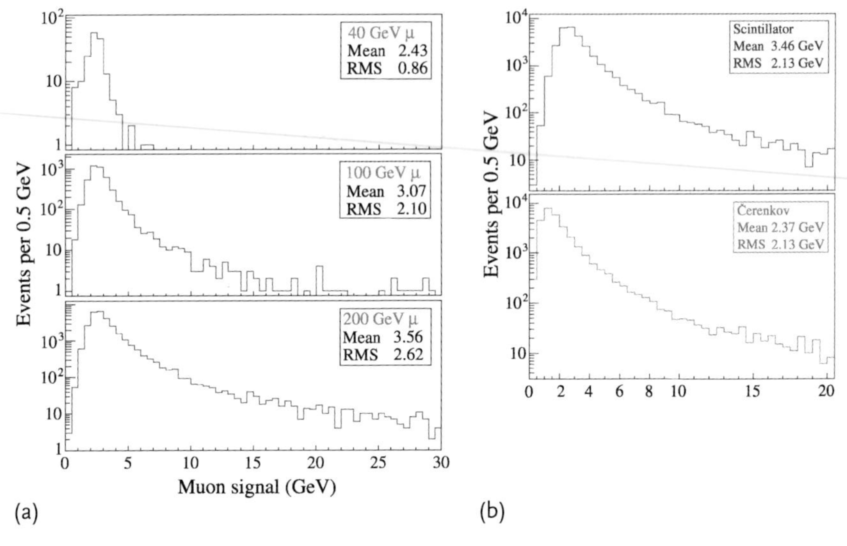

Figure 3.6 (a) The distributions of scintillation signals from μ^- traversing the 2-m-deep DREAM module at energies of 40, 100, and 200 GeV. The mean expected ionization energy loss is $\langle dE/dx \rangle = 1.1$ GeV, and the rapid increase in the radiative energy losses are obvious in these plots. The spectrum in energy displays the textbook dk/k distribution expected for a radiative process; and (b) a direct comparison of the scintillation and Čerenkov signals from 200 GeV μ^- traversing the DREAM module.

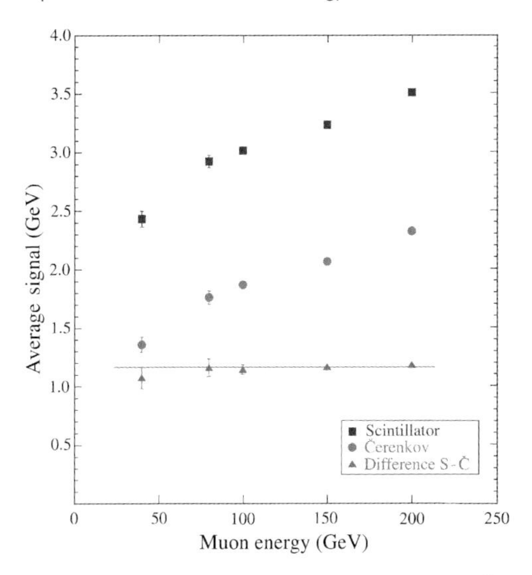

Figure 3.7 The mean scintillation (S) and the mean Čerenkov (C) response as a function of μ^- beam energy, and their difference, $(S - C)$, a constant at about 1.2 GeV.

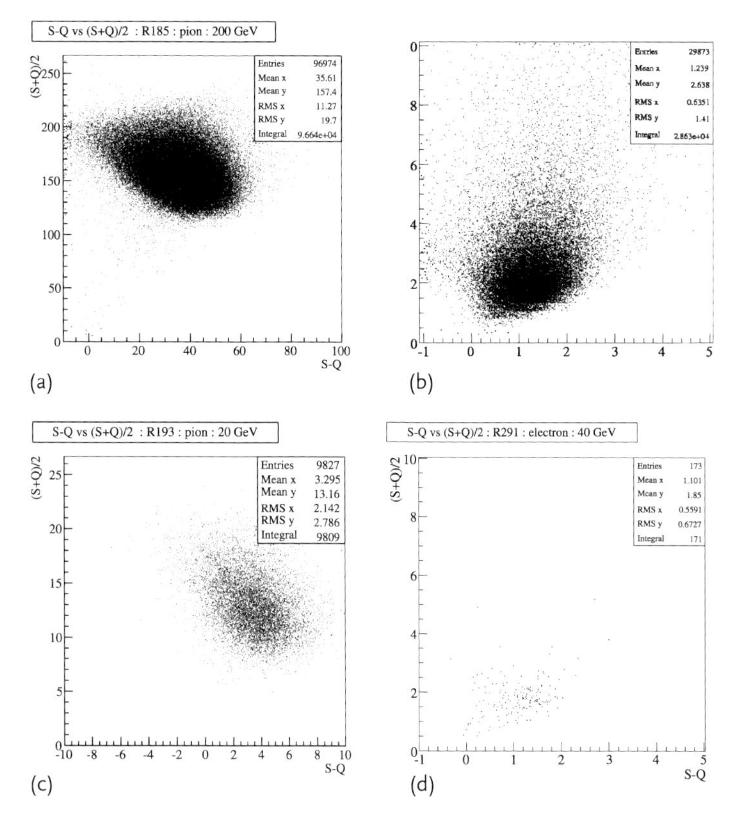

Figure 3.8 (a) The event-by-event sum of S and C signals, $(S + C)/2$, vs. the difference, $(S - C)$, for π^- beam at 200 GeV; (b) the same for μ^- beam at 200 GeV; (c) the average of S and C signals, $(S + C)/2$, vs. the difference, $(S - C)$, for π^- beam at 20 GeV; and (d) the same for μ^- beam at 40 GeV (substituted for 20 GeV).

The averaged signals in Figure 3.6 are not useful for individual events in an experiment, however. The same data ensembles have been used to plot directly, event by event, the measured quantities $(S - C)$ and $(S + C)/2$ for which we expect $\langle S - C \rangle \approx 1.2$ GeV and $\langle S + C \rangle/2 \approx E_{\text{radiated}}$. This is shown for 200-GeV π^- beam particles in Figure 3.8a and, correspondingly, for μ^- in Figure 3.8b. The 200-GeV μ^- are centered in $(S - C)$ around 1.24 GeV with an rms spread of 0.635 GeV. The mean in $(S + C)/2$ is about 2.64 GeV, with a radiative tail that extends up. In contrast, the 200-GeV π^- are far removed for the μ^- ensemble, allowing $\mu - \pi$ discrimination. The same plots at 20 GeV π^- and 40 GeV μ^- (substituted for 20 GeV μ^- where the beam flux was too low) are shown in Figure 3.8c and d.

A careful look at the numbers of events, about 100K π^- at 200 GeV and 10K π^- at 20 GeV, and the tiny overlap of the π and μ ensembles suggest an excellent μ-from-π discrimination at the 10^{-3}-to-10^{-4} level. In fact, the overlap in these beam data might be μ^- contamination from $\pi^- \to \mu^- \nu_\mu$ decay in flight in the π^- beam.

Nuclear particles interacting with nuclei (*A*) In a tracking system, in addition to the electromagnetic interactions with nuclei, hadronic particles ($\pi^{\pm}$, $K^{\pm}$, p, n) also interact in the materials of the detector and, in what may be a first, the CMS collaboration calculates the number of nuclear interaction lengths (λ_0), along with the number of radiation lengths (X_0), in its tracking system (Figure 6.13). This is a clear and honest acknowledgment that the huge jet backgrounds at the LHC, consisting predominantly of tens of $\pi^{\pm}$ and $K^{\pm}$ per jet, will have nuclear interaction inconsequences in the tracking detector.

In a calorimeter system, these hadronic interactions are overwhelmingly critical in hadronic calorimetry, which is largely dominated by the behavior of low energy particles, and the best resource for understanding the complexity and importance is the book *Calorimetry* by Wigmans [29] and discussed in Section 3.3.

3.2 Tracking Systems

Tracking detectors establish the existence of charged tracks, mostly from the interaction point but also from interactions and decays within the tracking volume. In a big detector, this volume can extend from tens of centimeters to two, or more, meters in radius, and includes the so-called vertex chamber close to the interaction point for the measurement of impact parameters. Since the total radial extent of the tracking plus calorimeter is usually limited by the radius of the solenoid, the choice of tracking volume vs. calorimeter volume, in both radial and axial dimensions, is a fundamental decision for a detector and reflects the goals of an experiment. Since the momentum resolution of a tracking system degrades at higher momenta proportional to p and improves with radial extent proportional to $1/R^2$, and the energy resolution of a calorimeter improves as $1/\sqrt{E}$ at higher energies provided it is deep enough, the choice of the radial and axial boundaries between the tracking system and the calorimeter system is driven by scientific goals, restricted by costs primarily in the calorimeter volume, and by engineering risks primarily in the solenoid. Given these cost and risk constraints, there will always be this fundamental conflict between tracking and calorimetry.

Some very well-known and successful experiments in the GeV-particle range, *for example*, Mark I at SLAC, operated essentially without a calorimeter, and later in the multi-GeV-particle range, *for example*, the L3 and OPAL experiments at LEP/CERN, operated with electromagnetic but not hadronic calorimeters. For future colliders, *for example*, ILC, CLIC, or the Muon Collider, it is accepted universally that any experiment will be equipped with very high performance electromagnetic and hadronic calorimeters. It is argued by some that the calorimeters should dominate the detector design and the tracking system be reduced to a charged-track sign and impact parameter measuring device. This is a minority view, but not crazy.

Momentum Resolution by Sagitta Measurement: The Sagitta Is Gaussian, Momentum Is Not

The sagitta, s, for an arc length, L, shown in the figure is related to the radius of curvature, R, as

$$s = R - \sqrt{R^2 - (L/2)^2} = R - R\sqrt{1 - (L/2R)^2} \approx \frac{L^2}{8R} ,$$

for $L \ll R$, and since the momentum, p, is related to the radius of curvature R in a magnetic field, B, as $p = 0.3\,BR$, the momentum and the inverse momentum of a track are

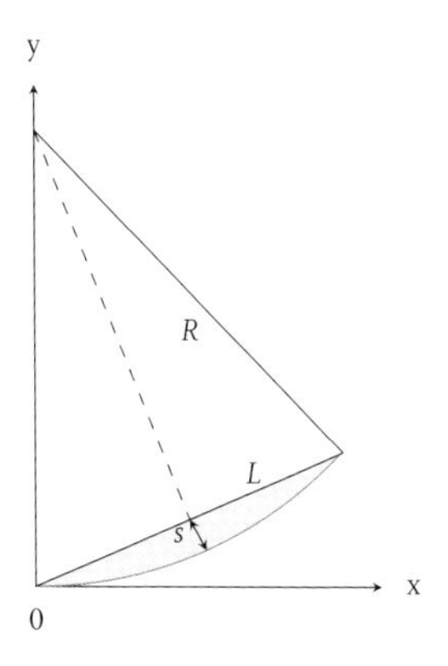

Figure 3.9 The sagitta of an arc of circle of radius R.

$$p = \frac{0.3\,BL^2}{8\,s} \text{ and } \frac{1}{p} = \left[\frac{8}{0.3\,BL^2}\right] s .$$

The primary measurements of the tracking system are presumed to be Gaussian distributed and perpendicular to the track, $x_i \pm \sigma_i$, from which the sagitta is calculated. Consequently, it is the inverse momentum, $1/p$, that is Gaussian distributed since $1/p \propto s$, and the momentum resolution can be found by taking differentials $\delta(1/p) = [8/(0.3\,BL^2)]\delta s$. An important result from Gluckstern [39] is that the optimum momentum resolution is obtained when one-fourth of the track measurements are at each end of the track and the remaining one-half in the middle, and for this geometry $s = (x_1 + x_4)/2 - (x_2 + x_3)/2$, so that $\delta s = \delta x_1/2 + \delta x_4/2 - \delta x_2/2 - \delta x_3/2$. For uncorrelated and independent spatial measurements, all cross-terms are zero, $\overline{\delta x_i \delta x_j} = 0$, and since the variance is defined as $\sigma^2 = \overline{(\delta x)^2}$, the rms sagitta resolution σ_s is given by

$$\sigma_s^2 = \overline{(\delta s^2)} = \overline{(\delta x_1/2)^2} + \overline{(\delta x_2/2)^2} + \overline{(\delta x_3/2)^2} + \overline{(\delta x_4/2)^2} = \sigma_x^2 ,$$

where σ_x is the average spatial resolution of each point. The differential of $1/p$ is $\delta p/p^2$, so the inverse momentum resolution can conveniently be written as

$$\frac{\sigma_p}{p^2}\,(\mathrm{GeV/c})^{-1} = \frac{8}{0.3\,B\,L^2}\sigma_x$$

A 1-m-long 100-GeV/c track in a 3 T field has a 1-mm sagitta, and zero sagitta is infinite momentum.

Therefore, momentum resolution improves linearly with spatial resolution and improves linearly with increasing B and quadratically with increasing L, and the resolution itself, σ_p/p, worsens linearly with momentum. There is no practical alternative to measuring momentum in this way, and large-scale experiments are pushing B toward 5 T and radii toward $L \sim 1.5$ m. A 1-m-long, 100-GeV/c track in a 3 T field has a 1-mm sagitta, and zero sagitta is infinite momentum.

The "Charpak" plot of tracking chamber performance over 100 years, Figure 3.13, displays a dramatic progression primarily due to an understanding of atoms. The advent of the multiwire proportional chamber (MWPC) and the drift chamber owed much to a careful understanding of the Townsend coefficients, the ionization of the electrons of atoms, and the movement of electrons and ions through gases [40]. The time projection chamber (TPC) [41], invented and developed by Nygren, built on the work of Charpak with a novel geometry, and the confidence to drift ionization electrons through 10 m of Ar gas.[24]

Tracking chambers now under active R&D are using atoms in even more novel ways: the "CluCou" drift chamber, based on the geometry of the KLOE chamber, clocks in every ionization cluster with a multi-GHz ADC and expects to achieve a spatial precision of 40–45 μm per wire and a specific ionization resolution for Poisson counted clusters of 3.5% [42]; and some have proposed sophisticated ideas for chambers that use not only the electronic properties of atoms but also their optical properties by measuring the optical light emitted by excited atoms to localize in three dimensions with high precision time and spatial measurements [43]. The Charpak plot would not be complete without track measurements in solid media, predominantly silicon [44].

3.2.1 Momentum Measurement of Charged Particles

The Lorentz force on a particle of momentum $\boldsymbol{p} = m\boldsymbol{v}$ and charge e in a magnetic field $\boldsymbol{B}$ equals the mass times the acceleration in a circle, $e\boldsymbol{v} \times \boldsymbol{B} = mv^2/r$, giving the useful relation

$$p = 0.3\,B\,r$$

24) The first TPC at SLAC/PEP4 drifted electrons 1 m through $Ar : CH_4$ gas at 8.5 atm.

in any of these combinations of units:

$$p\ (\mathrm{TeV/c}) = 0.3\ B\ (\mathrm{T})\, r\ (\mathrm{km})\,, \tag{3.4a}$$

$$p\ (\mathrm{GeV/c}) = 0.3\ B\ (\mathrm{T})\, r\ (\mathrm{m})\,, \tag{3.4b}$$

$$p\ (\mathrm{MeV/c}) = 0.3\ B\ (\mathrm{kG})\, r\ (\mathrm{cm})\,. \tag{3.4c}$$

Since the Lorentz force is perpendicular to the velocity, no work is done on, or by, the particle and it maintains a constant energy. In the bubble chamber photograph, Figure 2.3, the charged particles move in helices and lose energy (and momentum) mostly by ionization energy loss. A high energy track passing through this bubble chamber has small curvature, and tracking measurements are only on a limited arc length, ℓ, of the circle, as illustrated in Figure 3.9, from which the radius of curvature is derived in terms of the measured sagitta, s, as $r = \ell^2/8s$, so that the momentum resolution of a track depends on the spatial point precision as (Figure 3.9)

$$\frac{\sigma_p}{p} = \left[\frac{8\sigma_x}{0.3\ B\ell^2}\right] p\ (\mathrm{GeV/c})\,. \tag{3.5}$$

The "easy" way to achieve good momentum resolution is with ℓ^2, not B or σ_x, which can be costly. However, a calorimeter and a solenoid outside the tracking system will scale in volume like ℓ^3, and both of these are very costly.

The measured points on the arc have some spatial resolution, σ_x, which we assume to be Gaussian distributed. In a classic paper, Gluckstern [39] treats the momentum resolution of a tracking system in terms of point measurement resolution, multiple scattering contributions, including small angle nuclear elastic scattering, and the optimum spatial distribution of measured points. For discrete point measurements along a track, the optimum distribution is one-half of the measurements at the center and one-quarter of the measurements at each end.

More generally, a track will deviate from a straight line, as in Figure 3.10, for two reasons: the magnetic field bends it into a circle with a nonzero sagitta, and the nuclei of the medium will multiple scatter the track randomly side to side, resulting in a random sagitta whose rms width is

$$s_{\mathrm{rms}} = \frac{\ell}{4\sqrt{3}}\theta_{\mathrm{rms}}\,.$$

It is most convenient to work in terms of the curvature of the track,

$$k = \frac{1}{R} \quad \text{or} \quad k = 0.3\ B\frac{1}{p}\,,$$

or just $1/p$, in units of inverse momentum. The curvature uncertainty has two terms, $\delta k_{\mathrm{sagitta}}$ and δk_{ms}. For the important and practical case of N equally spaced

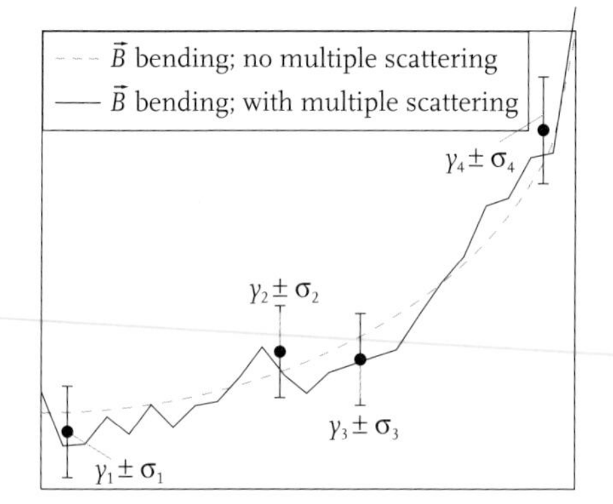

Figure 3.10 A sketch of a particle bending a magnetic field while multiple scattering in the medium. The sagitta rms, s_{rms}, contributes a random sagitta to the field bending and worsens the momentum resolution. If measurement point y_4 fluctuates up or down, the momentum (p) and azimuth angle (ϕ) of the track are negatively correlated.

points on a track each with spatial precision σ_x, the sagitta term (Gluckstern [39] and PDG [49, Section 28.12]) is

$$\boxed{\delta k_{\text{sagitta}} = \frac{\sigma_x}{\ell^2}\sqrt{\frac{720}{N+4}}}\,, \tag{3.6}$$

and if one end of the track is anchored at a vertex point, assuming the beam size is much smaller than σ_x, the number 720 under the square root becomes 320.

The multiple scattering sagitta, and therefore the curvature uncertainty due to multiple scattering, is

$$\delta\left(\frac{1}{R}\right) = \frac{8}{\ell^2}\delta s = \frac{8}{\ell^2}s_{\text{rms}} = \frac{8}{\ell}\frac{\theta_{\text{rms}}}{4\sqrt{3}}\,.$$

For a thin (in radiation lengths) chamber, the full expression for θ_{rms} should be used; the multiplication 8×0.0136 (GeV/c) $/4\sqrt{3}$ gives 0.0157 GeV/c, and so the final curvature uncertainty due to multiple scattering is

$$\boxed{\delta k_{\text{ms}} = \frac{0.0157\ \text{GeV/c}}{\ell p}\sqrt{\frac{\ell}{X_0}}\left[1 + 0.038\ln\left(\frac{\ell}{X_0}\right)\right]} \text{ or } \boxed{\delta k_{\text{ms}} \approx \frac{\theta_{\text{rms}}}{\ell}}\,.$$

A particle multiple scattering in a magnetic field is shown in Figure 3.10. The dotted line is the trajectory of the particle in a vacuum ($X_0 = \infty$), and the solid line is the actual spatial trajectory with multiple scattering. The four measurements, y_i, shown each have a Gaussian σ_i and fluctuate about the actual trajectory. In this particular illustration (with random numbers from PGFPLOT), there is no appreciable sagitta deviation due to multiple scattering. Since the inverse momentum is

$(1/p) = 0.3\,B(1/R)$ and the derivative is $\delta(1/p) = \delta p/p^2$, the momentum uncertainty, or resolution, is

$$\frac{\sigma_p}{p^2} = \frac{\sigma_x}{0.3B\ell^2}\sqrt{\frac{720}{N+4}} \oplus \frac{0.0157\,\text{GeV/c}}{0.3B\ell p}\sqrt{\frac{\ell}{X_0}}[1 + 0.038\ln(\ell/X_0)]\,. \qquad (3.7)$$

This general result has the familiar form

$$\frac{\sigma_p}{p^2} = [\text{ detector construction}] \oplus \frac{[\text{material budget}]}{p}\,,$$

where the "detector construction" means electronics, alignments, quantum fluctuations that contribute to σ_x, and so on, and "material budget" means essentially the radiation lengths of matter that the particle traverses. The material budget "doesn't matter" for high momentum tracks, which is true enough, but a precision experiment should strive to measure all momenta well, even down to 1 GeV/c.

In addition, a forward track will lose measurements in the tracking system because it does not reach the full tracking radius, and this loss in number of measured points goes like $\sin^{1/2}\theta$. Another factor of $\sin\theta$ is lost in $\boldsymbol{F} = e\boldsymbol{v} \times \boldsymbol{B}$, and another in the projection of the transverse momentum. Overall, the momentum resolution as a function of θ as $\theta \to 0$ scales as

$$\frac{\sigma_p}{p^2} \sim \frac{1}{\sin^{5/2}\theta}\,, \qquad (3.8)$$

and this is geometry only. Everything about the detector is harder in the forward direction since the intense beams are nearby and the services (power in and signals out) are also often crowded into this region. Since the loss can be large and there is good physics to be done in the forward region, so-called "forward tracking" has been identified (Chris Damerell) [45] as a critical and unsolved part of a big detector.[25]

3.2.2
Impact Parameter Measurement

A three-dimensional figure of the D0 vertex chamber is shown in Figure 3.11 with its "barrel disks" and the two sets of axial, or "end cap", disks. Most vertex chambers have a similar geometry, and the end cap disks are important for the reasons given above, but also because D0 operates at the $p\bar{p}$ Tevatron Collider where there is a large flux of particles from diffractive processes at low angles. A very recent event recorded by the ATLAS vertex chamber triggered on a cosmic muon is shown in Figure 3.12. Current design work in the context of the ILC is noted in the review [46].

25) On the 4th (Chapter 7) we played with the idea of a forward tracking toroid, in the spirit of ATLAS, but inside the calorimeter. The numbers for superconducting currents and the necessary sagitta measurements are not easy.

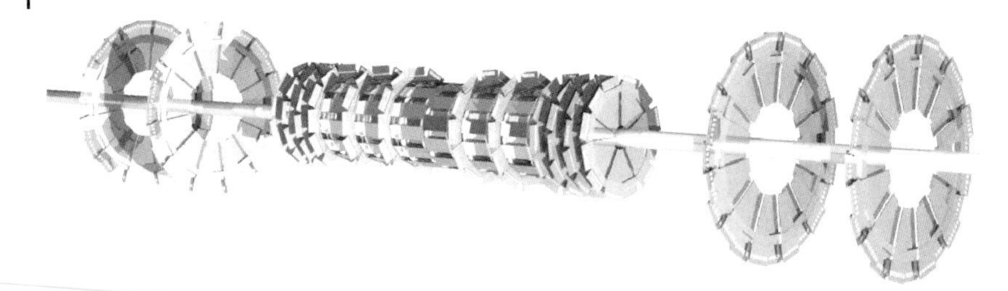

Figure 3.11 D0 silicon strip vertex chamber at the Fermilab Tevatron Collider. Note the central barrel region with silicon wafers on azimuthal cylinders, and the two sets of axial disks, one set of two at larger z. This geometry is optimum for the physics and backgrounds at a $p\bar{p}$ collider at $\sqrt{s} = 2$ TeV.

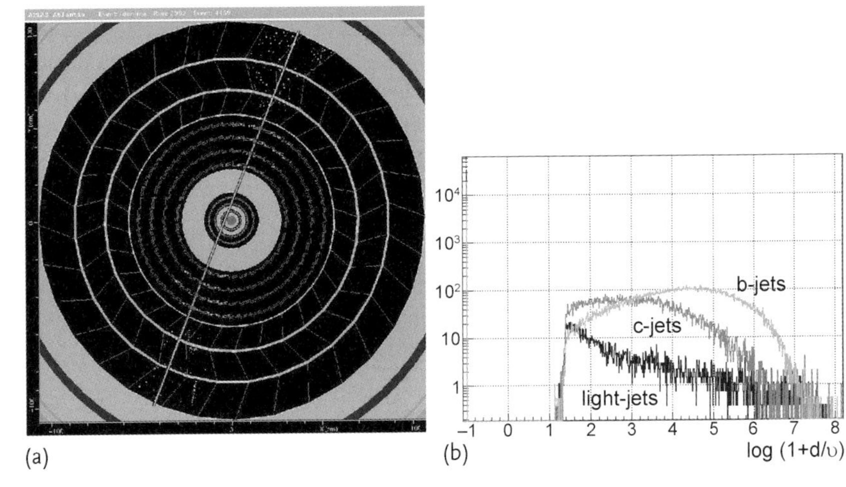

Figure 3.12 (a) A cosmic muon recorded in the ATLAS detector. The actual spatial precision is much smaller that the size of the marks in this reconstruction. (b) Simulated results for the impact parameter tagging significance, plotted as $\log_{10}(1 + b/\sigma_b)$, for three ensembles of jets: c-jets, b-jets, and uds-jets.

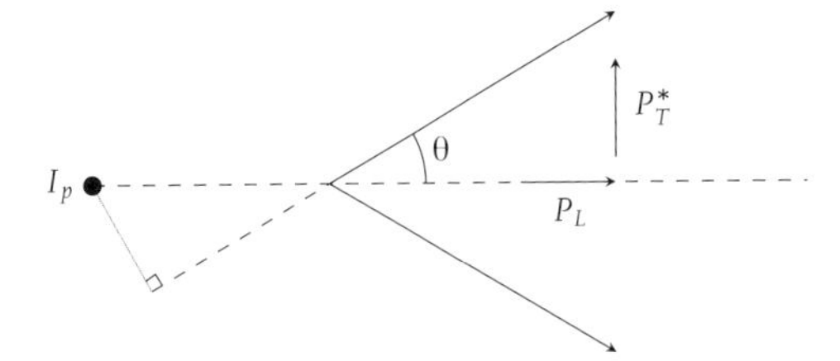

The measurement of an impact parameter, b, and the expected resolution on it is pure geometry, but it is not simple. There are two effects, the first a geometrical extrapolation from measured points on the track to the (very small) beam crossing region (or interaction diamond), the second multiple scattering effects in the beam pipe material and the vertex chamber itself. The impact parameter can be estimated from a least-squares fit to the trajectory (using the curvature information from the overall track fit), and the resolution on b will depend linearly on the point resolution σ_x of the chamber. For r_1 the radius of the first chamber layer, r_N the radius of the last layer, and measured points in between y_i, $i = 1, N$, the resolution in b goes like

$$\sigma_b^x \approx \xi \sigma_x\,, \quad \text{where} \quad \xi \sim \frac{r_1}{r_N - r_1}\frac{1}{\sqrt{N}}\,,$$

that is, ξ is a function only of the number and radii of the chamber layers.

The multiple scattering contribution is driven by the basic rms angular spread of a particle of momentum p traversing material of thickness ℓ/X_0 in radiation lengths,

$$\theta_{\text{rms}} = \frac{0.0136\,\text{GeV/c}}{p\beta c}\sqrt{\frac{\ell}{X_0}}[1 + 0.038\ln(\ell/X_0)]\,.$$

The term involving the logarithm is an approximation meant to account for single and plural Coulomb scattering in thin absorbers where the Central Limit Theorem is not fully satisfied; it is important for vertex chambers, which are thin. The first consequence is directly the deviation from a zero-impact parameter due to θ_{rms} over the lever arm r_1, the radial distance to the first layer (or first piece of material) of the vertex chamber, which is $\sigma_b^{\text{ms}} \approx r_1\theta_{\text{rms}}$. For a track at polar angle θ with respect to the beam, this expression is augmented in two ways: first because the material depth increases, $\ell \to \ell/\sin\theta$, and second because the level arm is larger, $r_1 \to r_1/\sin\theta$. This results in

$$\sigma_b^{\text{ms}} \approx \frac{r_1}{\sin\theta}\frac{0.0136\,\text{GeV/c}}{p\beta}\sqrt{\frac{\ell/\sin\theta}{X_0}}\left[1 + 0.038\ln\left(\frac{\ell/\sin\theta}{X_0}\right)\right]\,.$$

Gathering the $\sin\theta$ and ln terms into ζ as

$$\zeta = \frac{r_1}{\sin^{3/2}\theta}[1 + 0.038\ln(1/\sin\theta)]\,,$$

the contribution to σ_b can be simply written

$$\sigma_b^{\text{ms}} \approx \zeta\theta_{\text{rms}}\,.$$

The third contribution is from the lateral displacement of a track, $y_{\text{rms}} = r\theta_{\text{rms}}/\sqrt{3} \sim 1/p$, which is a contribution from the several layers of the chamber, so summing in quadrature results in an impact parameter contribution that looks like $\sigma_b \approx \eta/\sqrt{p}$. The overall resolution on b is a sum in quadrature of this term, plus σ_b^x and σ_b^{ms}, but care must be taken for correlations among these terms, as cautioned in PDG [49, Section 27.3]. Ignoring these correlations (which means that this problem should really be simulated to get it right), the resolution on b is the sum in quadrature of these three terms:

$$\sigma_b \approx (\xi\sigma_x) \oplus (\zeta\theta_{\text{rms}}) \oplus (\eta/\sqrt{p})\,. \tag{3.9}$$

Therefore, a better impact parameter resolution is achieved for small r_1 (get close to the beam) and small $\sqrt{\ell/X_0}$ (reduce the material in the chamber), and at a higher momentum everything is better.

A goal for a future detector is to achieve numbers like the following:

$$\sigma_b \approx 5\ \mu\text{m} \oplus 10\ \mu\text{m}/p \sin^{3/2}\theta \oplus 10\ \mu\text{m}/\sqrt{p} \tag{3.10}$$

with pixels of approximate area $(20\ \mu\text{m})^2$. Comparing the expressions in Eq. (3.9) with the numerical goals in Eq. (3.10) implies daunting technological breakthroughs. Several groups are working on several different technologies as discussed by Damerell [46] and Brau [47].

For approximately these numbers for σ_b, the significance for physics of impact parameter tagging in the reconstruction of top quarks in the physics process $e^+e^- \to t\bar{t} \to 6-$ jets at a $\sqrt{s} = 500$ GeV linear collider, the impact parameter tagging significance for c-jets, b-jets, and uds-jets is shown[26] in Figure 3.12b. Expressed in the figure as $\log_{10}(1 + b/\sigma_b)$, it is clear that b-quarks inside of B mesons with $c\tau \sim 500\ \mu$m populate the region of high significance, and the c-quarks inside D mesons with $c\tau \sim 100-300\ \mu$m populate the middle region of significance. The light quarks, uds, are peaked at zero, since almost all mesons and baryons with uds quarks inside of them have $c\tau$ in the millimeter or larger range. The high-side tail on the uds significance may be due to the multiple scattering term in σ_b, $(\zeta\,\theta_{\text{rms}})$, or to a small number of short-lifetime strange particle decays.

Finally, having a vertex chamber out to $R \sim 0.2$ m followed by a tracking chamber out to $R \sim 1.5$ m, both of which measure charged tracks that have been presented here with different detector functions (momentum resolution vs. impact parameter resolution), can their functions be combined, for example, for improving the momentum or impact parameter resolution? The answer is "yes" if the multiple scattering is small, in which case they can function as one large tracking chamber with an effective track arc length of $\ell \sim R$. However, if there is material for supports or containment between the vertex chamber and the tracking, multiple scattering in this material will decouple the two chambers and, for example, the momentum measurement will not be able to take advantage of the longer arc length, $\ell \sim R$. This is treated in the formulation by Gluckstern [39].

3.2.3
Summary of Tracking

1. The resolution of the inverse momentum is Gaussian, not the momentum itself, which is divergent at high momentum as the sagitta goes to zero. This derives from the Gaussian measurement of the sagitta of the track.
2. The optimum (for momentum resolution) distribution of measured points on a track is one-half the points at the center of the track, and one-quarter at each end.

26) Solution by Fedor Ignatov, Budker Institute, and described in detail in the Letter of Intent, http://www.4thconcept.org/4LoI.pdf.

3. The momentum and the azimuth angle are negatively correlated, that is, $\overline{\delta p \delta \phi} < 0$.
4. For helical tracks with pitch angle λ (sometimes called a dip angle from the center plane of the detector, or polar angles $\theta \to 0$ or $\theta \to \pi$), only the component of momentum p perpendicular to B contributes to the sagitta, and in addition the depth of material is larger, making everything worse. In particular:

$$\delta k_{\text{sagitta}} \to \delta k_{\text{sagitta}}/\cos^2\lambda \quad \text{and} \quad \delta k_{\text{ms}} \to \delta k_{\text{ms}}/\cos^2\lambda \, .$$

5. Modern methods for track finding from discrete point measurements, such as a Kalman filter[27] do not alter Gluckstern's results on resolutions.

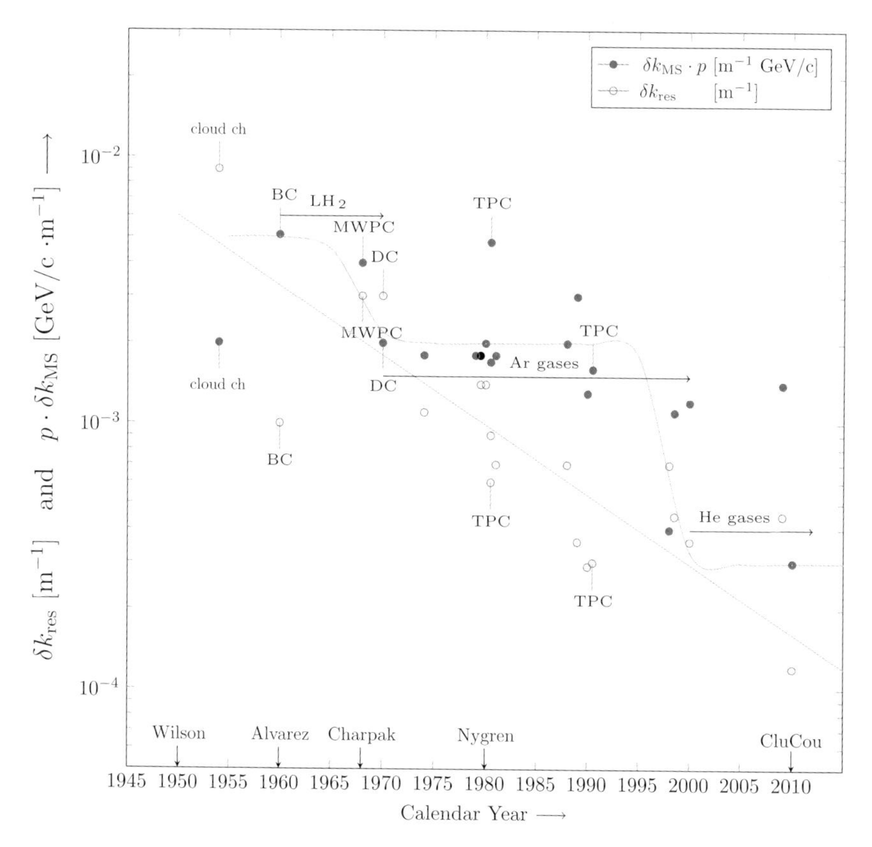

Figure 3.13 The curvature uncertainties due to spatial resolution fluctuations (○) $\delta k_{\text{res}} \sim \sigma_x/\ell^2 \cdot \sqrt{720/(N+4)}$, and due to multiple scattering (•) $\delta k_{\text{MS}} \sim \theta_{\text{rms}}/\ell$ at $p = 1\,\text{GeV}/c$ in the materials of a given tracking chamber of length $\ell = 1\,\text{m}$ as a function of calendar year for existing chambers. The total curvature uncertainty of a track is $\delta k = \sqrt{(\delta k_{\text{res}})^2 + (\delta k_{\text{MS}})^2}$. One line shows the direct improvements in resolution, the other shows the step-wise reductions in chamber material. CluCou relies on cluster timing in a light gas mixture (He) proposed by Grancagnolo. [42]

27) P. Billoir in the DELPHI experiment at LEP (Comp. Phys. Comm. 1989).

6. The impact parameters from a two-body decay of a particle are approximately independent of the momentum of the particle.

Progress over 60 years in tracking chamber resolutions is shown in Figure 3.13.

3.3
Calorimetry: Energy Measurement

"Everything about calorimeters is obvious, once you understand it."
– Richard Wigmans

A calorimeter is designed for the total absorption of a particle's energy, that is, a particle enters, interacts, its secondaries reinteract, and so on, until all byproducts are reduced by dE/dx to zero energy. The calorimeter medium is interspersed with charge-sensitive detectors (a "sampling" calorimeter) or the medium itself is sensitive to charged particles (a "homogeneous" calorimeter). Clearly, the calorimeter is placed after the tracking system and is the "detector of last resort" for all particles except $\mu^{\pm}$.

Calorimeters from the 1960s to the 2010s have progressed from "shower counters" (e.g., Magnetic Detector, or Mark I), to the excellent ZEUS calorimeters, to sophisticated calorimeter designs intended for detectors at TeV particle energies [48] with very high precision. Currently, calorimeters exceed tracking chambers for energy (or momentum) measurements at the highest energies. The energy resolution of a calorimeter, σ_E/E, improves at higher energies, while the momentum resolution of any sagitta-measuring tracking system degrades linearly with momentum, as shown in Figure 3.23 for the best expected future tracking chambers and calorimeters. It will be a supreme challenge to attain these numbers in the foreseeable future. Given these difficulties, there have been many proposals to build big detectors that rely almost completely on calorimetry and relegate the tracking system to a small volume that merely (i) points charged tracks at the calorimeter without even a magnetic field (*for example,* D0 at the Tevatron Collider in run 1, (ii) measures only the sign of charged tracks, (i.e., tags μ^+ and μ^-, etc. to reduce combinatorial backgrounds in physics ensembles), or (iii) tags impact parameters to high precision. It is not clear that any one of these half-measures is good enough for a comprehensive no-fail collider detector.

It is now generally accepted that calorimeters will not only be essential in future colliders, but must also measure jets to match the precisions on electrons and muons. This challenge has not yet been met in a large system. For this basic reason, a good understanding of calorimeters is critical to all new collider detectors. The best reference on calorimetry of all kinds is by Wigmans, *Calorimetry: Energy Measurement in Particle Physics* [29].

Calorimeters are expensive by the cubic meter, and the volume in a collider detector is always large: if the tracking system ends at $R \sim 1.5$ m radially and $L \sim 2.0$ m axially, and the calorimeter is 1.5 m deep, its volume is 150 m^3 and a typical cost is \$1 M/m^3. Total absorption means the density is high, say 6 g/cm^3, so the mass is approximately 1000 metric tonnes, by far the most massive item in the detector. Cost considerations imply a smaller tracker radius since the cost is proportional to R^3, but then momentum resolution is degraded slightly faster than $1/R^2$ (since the arc length ℓ is less than R). Making the calorimeter thinner than 1.5 m (and perhaps trying to increase the density to compensate) leads to leakage and a degradation of energy resolution. There is no easy choice, and cost optimization studies do not have strong minima.

Even overall physics performance optimizations do not have strong maxima.

Electromagnetic Calorimeter Energy Resolution Scales as $\frac{\sigma_E}{E} \sim \frac{k}{\sqrt{E}}$

The critical energy, E_c, of any material is approximately $E_c \approx X_0(\frac{dE}{dx})_{\text{min}}$, and this energy drives the character of an EM calorimeter. For energies above E_c, particle production through bremsstrahlung ($e \to e\gamma$) and pair production ($\gamma \to e^+e^-$) in the field of high-Z nuclei proceeds vigorously and the electromagnetic particle population increases exponentially in depth (a doubling every X_0). Below E_c ionization energy loss dominates and the shower is depleted. The maximum in particle population is reached when the average energy of the shower particles is E_c and $N_{\text{max}} \approx E/E_c$. N_{max} is Poisson with an expected fluctuation of $\sigma_N \approx \sqrt{N_{\text{max}}}$ and, since the total energy is proportional to N_{max}, the energy resolution is $\sigma_E/E \approx k/\sqrt{N}$, where N is the measured number of shower particles (not including, for example, the γs), and E is proportional to N, so that

$$\frac{\sigma_E}{E} = \frac{k}{\sqrt{E}} .$$

A complete treatment of electromagnetic showers is given by Wigmans [29, Chaps. 2, 7] including often overlooked subtleties, *for example*, that the shower particles are mostly high energy *mips* in the early part of a shower, and mostly Compton electrons from the numerous MeV-γs in the later parts of the shower, and these have detector responses that can differ by 30%.

3.3.1 EM Particles, $e^\pm$ and γ

Electromagnetic showers in any medium are simple to understand since an electron traversing atoms will emit photons by bremsstrahlung in the intense electric field of a nucleus $+Ze$, $e \to e\gamma$, and a photon will pair-produce e^+e^- in the same field, $\gamma \to e^+e^-$. The electrons in the atoms with charge $-1e$ have a negligible influence on these processes, which depend strongly on the nuclear charge ($+Ze$) of the medium. In both processes, the nucleus is left intact and ensures momentum-energy conservation while absorbing negligible momentum and energy itself. The spatial scale for this EM-particle doubling is a "radiation length", X_0, which depends on Ze. Roughly, the radiation length of a material is

$$X_0\,(\mathrm{g/cm^2}) \approx \frac{716.4\,\mathrm{g\,cm^{-2}}\,A}{Z(Z+1)\ln(287/\sqrt{Z})} \approx \frac{180}{Z} \tag{3.11}$$

from a fit by Dahl [49]; for four-digit precision, see PDG [49]. Physically, the radiation length is the distance over which the average electron energy is reduced to $1/e$ of its original energy by bremsstrahlung, *and* it is (7/9) of the mean free path of a high energy photon.[28] These properties for several common metals are listed in Table 3.1.

This doubling of the EM particle population every X_0 results directly in an exponential increase in the number of e and γ shower particles, which continues for the $e^{\pm}$ until they are degraded in energy to the "critical energy" of the material, Figure 3.14, at which an $e^{\pm}$ is more likely to lose energy by dE/dx rather than further particle production. The γs of the shower cease pair-production of e^+e^- at the kinematic threshold, $E_\gamma = 2m_e c^2 \approx 1\,\mathrm{MeV}$, but continue to deposit substantial energy through Compton scattering at sub-MeV energies. This has important implications for calibration of an EM calorimeter that is segmented in depth [29, Section 2.2.4].

The calorimeter energy measurement is essentially a count of the number of $e^{\pm}$ in the shower, which, for incident particle energy E, is approximately

$$N \approx E/E_c\,,$$

Table 3.1 Physical properties of calorimeter absorbers; adapted from Wigmans [29, App. B].

Absorber material	Z	A	Density (g/cm³)	E_c (MeV)	X_0 (mm)	λ_{int} (mm)	$(dE/dx)_m$ (MeV/cm)
C	6	12.0	2.3	83	188.0	381	4.0
Al	13	27.0	2.7	43	89.0	390	4.4
Fe	26	39.9	7.9	22	17.6	168	11.4
Cu	29	63.6	9.0	20	14.3	151	12.6
W	74	183.9	19.3	8.0	3.5	96	22.1
Pb	82	207.2	11.3	7.4	5.6	170	12.7
^{238}U	92	238.0	18.9	6.8	3.2	105	20.5
concrete	—		2.5	55	107	400	4.3
glass	—		2.2	51	127	438	3.8

28) That is, a photon has a mean free path of $(9/7)X_0$, and this has implications for $\gamma - e$ shower development differences in EM calorimeters [29].

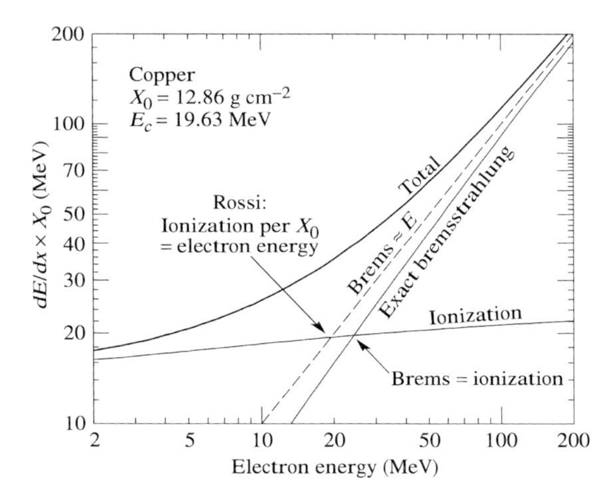

Figure 3.14 The critical energy, E_c, of a material is the crossover energy between energy loss by ionization and energy loss by radiation. The vertical axis is one estimate of the critical energy, $E_c \approx X_0 dE/dx$. The difference in definitions is not important, except for high precision simulations.

and N is a Poisson number with an expected statistical fluctuation of $\sqrt{N}$, and therefore the energy resolution must scale as

$$\frac{\sigma_E}{E} \approx \frac{\sigma_N}{N} \approx \frac{k}{\sqrt{E/E_c}},$$

or

$$\sqrt{E}\left(\frac{\sigma_E}{E}\right) = k\sqrt{E_c},$$

where k is a constant that depends only upon the geometry and construction of the calorimeter, and E_c that depends only on the absorber medium. An approximately universal plot of many EM calorimeters of various sampling frequencies, f_{samp}, and sampling spatial granularity, d(mm), is shown in Figure 3.15 plotted as

$$\sqrt{E}\left(\frac{\sigma_E}{E}\right) = 2.7\%\sqrt{\frac{d\ (\text{mm})}{f_{\text{samp}}}}.$$

The E_c variation among these calorimeters is small since all are constructed from high-Z absorbers. In general, electromagnetic calorimeters are "easy", and it is difficult to make one that is nonlinear or with poor energy resolution.

3.3.2 Hadronic Particles (Particles Composed of Quarks)

Hadronic calorimeters must cope with a much more varied collection of possible reactions and are not easy to understand. The overall dependence on energy is,

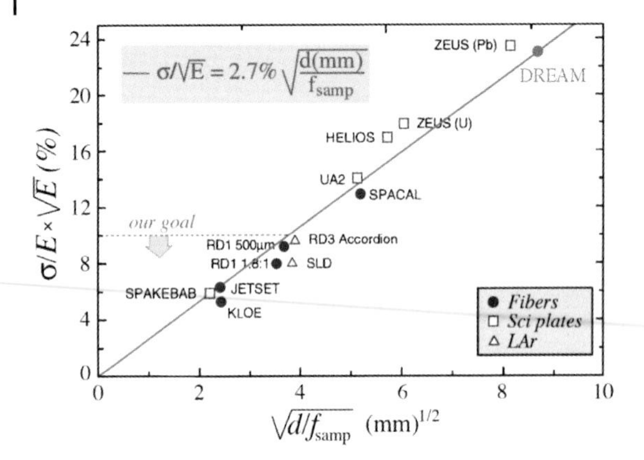

Figure 3.15 Electromagnetic calorimeters fall along an easily understood curve. This is an updated plot originally published in Wigmans [29, Figure 4.8].

grosso modo, still given by an expression of the form $\sigma_E/E \sim k_{\text{had}}/\sqrt{E/E_{\text{had-c}}}$ for $E_{\text{had-c}} \sim 1\,\text{GeV}$ (approximately a multipion production threshold), but this only masks the complexity of hadronic showers. The four events shown in Figure 3.16 are all sequential events recorded in a sampling calorimeter with readout every sampling layer in depth. The large localized energy deposits are $\pi^0 \to \gamma\gamma$ and are produced in the hadronic cascade and immediately decay into two photons. A close examination shows that the event-to-event fluctuations in hadronic showers are so huge, in both relative and absolute senses, that the total energy deposit as recorded depends on the amount of electromagnetic energy in the shower, that is, the number of π^0 produced relative to the number of $\pi^\pm$ produced, and that penetration of hadronic particles persists in depth. All of these characteristics are completely different from those of electromagnetic showers in which, essentially, every shower is the same.

A brief life history of a hadronic shower begins with a high energy hadron, for example a $(u\bar{d})$ charged pion, π^+, entering the calorimeter volume. Apart from minor dE/dx ionization, the π^+ will strike a nucleus, often ejecting an energetic nucleon as in ordinary $\pi^+ p$ interactions and itself diffractively producing forward-going energetic pions. The scattered nucleon may rescatter on other nucleons, there may be some energy deposit into the surviving nucleus, or the nucleus may disassemble into nucleons and some nuclear fragments such as α particles. The protons repel each other, gaining kinetic energies of 100 MeV; the neutrons are left with their Fermi energies of a few MeV, and if a nucleus of any kind remains, it may deexcite by repeated γ emissions. The incident π^+ has given up its energy in the collision, including energy to overcome the binding energies of the nucleons. The ratio of neutrons to protons depends on A, Z. The secondary hadrons will continue to break up additional nuclei. These secondaries, and further generations of particles, are predominantly pions, π^+, π^-, π^0 in approximately equal numbers, plus of order 10% kaons, $K^+, K^-, K^0, \bar{K}^0$, and a small number of nucleons, n, p from the hadron-nucleus collision. The neutral pions decay immediately

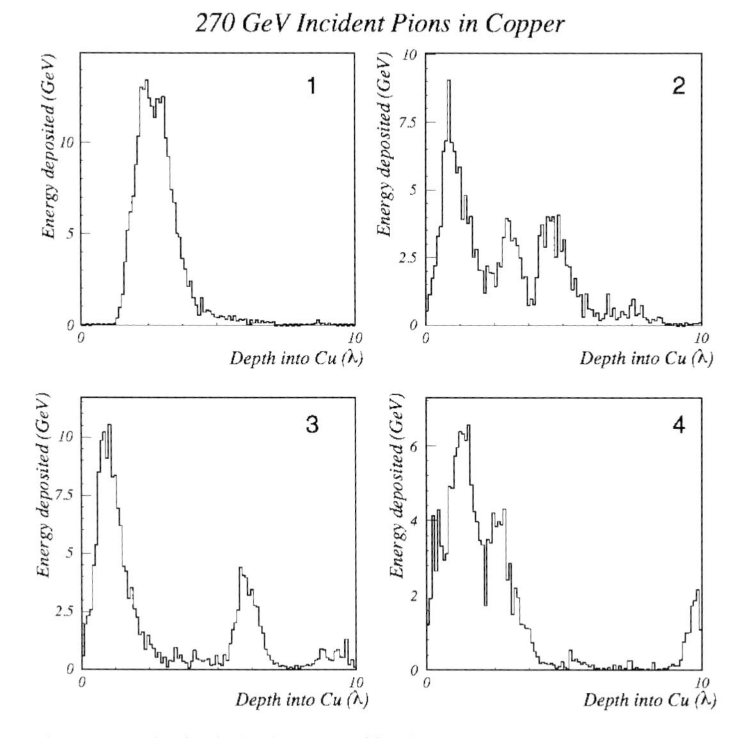

Figure 3.16 The depth development of four hadronic showers, each initiated by a 270-GeV π^-. The large localized energy deposits are π^0s produced in the calorimeter volume and decaying immediately as $\pi^0 \to \gamma\gamma$.

($\tau \approx 10^{-16}$ s) into $\pi^0 \to \gamma\gamma$, which become individual EM showers. Therefore, through π^0 production, a fraction of the hadronic energy is converted completely into electromagnetic energy in the calorimeter. The inverse does not happen (electroproduction and photoproduction cross-sections are very small), and therefore there is a net transfer from hadronic to electromagnetic energy, and the mean of this transferred energy grows with shower energy. Poisson fluctuations in the number of produced π^0 cause fluctuations in the electromagnetic content, or EM fraction (f_{em}), in hadronic showers. Since electromagnetic showers generate many charged $e^{\pm}$ tracks, which in general give a large signal compared to $\pi^{\pm}$, the electromagnetic response (e) is larger than the hadronic response (h), written as $(e/h) > 1$, and called "noncompensating". A "compensating" calorimeter has $e = h$.

The incident π^+ was noted in the above brief history to be a $(u\bar{d})$ state for a good reason. An incident proton (uud) behaves quite differently in a calorimeter for reasons that can be loosely attributed to baryon conservation. A proton is less likely to break up, or if it does, then it tends to remain a baryon, say a neutron (udd). This leads to some distinct differences in calorimeter response, in both the mean response and the resolution [50].

The best calorimeter ever built and tested was SPACAL [51], and the best calorimeters ever used in an experiment were the calorimeters of the ZEUS experiment at

HERA. In both of these, careful attention was paid to compensation effects. For this reason, the next generation of calorimeters at big colliders will almost certainly be "dual-readout" calorimeters that maintain all the advantages of compensation but with great flexibility in the absorber and sensor media.

For illustration, a small sample of energy distributions [51] from SPACAL at π^- beam energies of 10, 40, and 150 GeV are shown in Figure 3.17. These distributions are symmetric and Gaussian, even far into the top-side and bottom-side tails, and they are narrow with direct energy resolutions of 11.0% at 10 GeV, 5.75% at 40 GeV, and 3.46% at 150 GeV. One can, of course, represent these resolutions in the form $\sigma/E = k/\sqrt{E} \oplus c$ and quote k and c to represent the calorimeter, but it is better to just quote the resolution at each energy. It is the resolution that matters in a physics problem, and it does not matter if these energy fluctuations arose from imperfections in the detector or from quantum fluctuations in the signal generation.

I argue that these plots, and the many more like them from the SPACAL calorimeter, should be the standard against which all future proposed calorimeters are compared.

Currently running experiments at hadron colliders (CDF, D0, ATLAS, CMS) display a variety of hadronic calorimeter technologies with widely differing strengths and weaknesses. None of them is capable of achieving a clear signal for $W^\pm \to jj$ or $Z^0 \to jj$ in a direct plot of two-jet mass resolution due to their poor hadronic energy resolution on the two jets. For small jet masses, the invariant mass squared of a $W^\pm$ or Z^0 is

$$M^2 \approx 2E_1E_2(1 - \cos\theta_{12}) \approx E_1E_2\theta_{12}^2 ,$$

and in order to achieve a W, Z mass resolution comparable to their natural widths, 2–3 GeV, a jet energy resolution of about

$$\frac{\sigma_E}{E} \approx \frac{30\%}{\sqrt{E}} \oplus 0.5\%$$

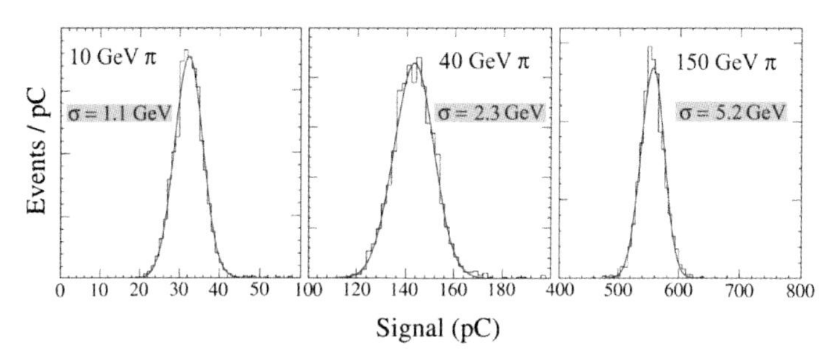

Figure 3.17 Raw pulse height distributions (in pC) from the compensating calorimeter test module SPACAL exposed to π^- beams of 10,40, and 150 GeV. The quality and fidelity of these calorimeter measurements are evident from the Gaussian line shape, the absence of tails on both the high side and the low side, and the narrow (fitted) rms widths shown in the figure for each energy. We argue that a scientific assessment of calorimeter performance is best obtained by looking at these response functions in their full glory. A quoted "rms" or a partial fit are not sufficient to judge calorimeter quality.

is required, that is, a stochastic term of 30% and a constant term that is small, of order 0.5%. These two numbers are very difficult to achieve in a working calorimeter, especially in a large system that is integrated into a large detector. The anticipated future lepton lepton colliders, either e^+e^- or $\mu^+\mu^-$, will be required to achieve these two numbers explicitly so that $W^\pm \to jj$ and $Z^0 \to jj$ decays will be useful for physics, not to mention calibration. One might say that the importance of energy resolution to a detector is like the cube: one σ_E for mass definition for physics, one σ_E for background rejection, and one σ_E for calibration in determining the mean of a distribution to calibrate energy. Thus, the overall detector impact is proportional to σ_E^3, and every 10% in σ_E is important for a detector.

There are only two generic options for hadronic calorimetry presently under study, particle flow analysis (PFA) calorimeters and dual-readout calorimeters. Between these, there are many variations and options for absorber materials, photosensors, digital or analog, absorber depth with or without tail catchers, and several sensitive media. For dual-readout calorimeters, sensitive media of fibers, crystals, and combinations have been built and tested [52–55], and geometries with planar alternating Čerenkov and scintillator plates as dual-readout media are planned.[29] There are many options within the crystal/glass category alone, and the optical issues of attenuation, refraction, photoconversion, radiation damage, and so on are not fully explored. It is a rich and vital area of R&D for any future collider.

Dual-Readout Calorimetry

Long ago the principle of compensation ($e/h = 1$) in hadronic calorimeters was demonstrated [51], predictions were made for the performance of not-yet-built calorimeters that were borne out by beam test measurements, and a hadronic energy resolution of $30\%/\sqrt{E}$ was achieved with a small constant term. This calorimeter was SPACAL, a 9-t Pb plastic scintillator (4:1 ratio) built out of longitudinal fibers embedded in a Pb matrix. Compensation means that the energy response of the calorimeter to electrons (e) is the same as its response to hadrons (h) at all energies. This is unusual, since in almost all calorimeters the electromagnetic response is larger, $e/h > 1$, sometimes by large factors. In the SPACAL configuration, compensation was achieved by boosting the hadronic signal by measuring the neutrons liberated in nuclear breakup, that is, by recovering the binding energy losses incurred in the hadronic shower, and the mechanism was to measure the proton recoils from the neutron elastic scatters ($np \to np$) in the plastic scintillating fibers. Because the neutrons are liberated with their Fermi energies $T_n \approx 2$–5 MeV, the neutron velocity in the calorimeter is $\beta \approx \sqrt{2T_n/m_n c^2} \sim 0.06$. Therefore, the neutrons linger in the calorimeter volume long after the hadronic shower is depleted. In the DREAM module [56] the neutron mean free path is $\lambda_n = 56$ cm at $E_n = 3$ MeV and $\lambda_n = 10$ cm at $E_n = 0.1$ MeV. At both ends of this energy range, the mean time between collisions is 23 ns. Consequently, the

29) For example, the Homogeneous Hadronic Calorimeter (HHCAL) meeting, IHEP, Beijing, 9 May 2010, in coordination with the CALOR10 conference.

recoil signals come later than the relativistic shower development, out to times of 100 ns, and fill a larger volume, out to 1 m. Thus, the SPACAL module only achieved an energy resolution near $30\%/\sqrt{E}$ after integrating out to 100 ns and including signals from the whole 9-t module.

Dual-readout can be (and was originally conceived to be [57]) thought of as "flexible compensation". There is no need to build into the instrument a fixed absorber-sensor compensating ratio, *for example*, 4:1 for a Pb scintillator. Instead, one can measure both parts of a hadronic shower, the electromagnetic part (essentially the produced $\pi^0 \to \gamma\gamma$) and the hadronic part (essentially everything else). You can use any absorber and any sensor (within reason) in any ratio you like. In the geometry of an optical fiber calorimeter such as DREAM, the electromagnetic part that predominantly consists of subshowers from $\pi^0 \to \gamma\gamma$ decays is directly measured by the Čerenkov light generated by the relativistic $e^{\pm}$ passing through the clear fibers (both pure quartz and inexpensive plastic optical fibers were used in the DREAM module). The scintillating fibers measure the dE/dx contributions from all charged particles, the $e^{\pm}$ plus all the nuclear parts ($\pi^{\pm}$, $K^{\pm}$, spallation p, recoil p, nuclear fragments, etc.). A simple and direct plot of these two signals is shown in Figure 3.18 with the total Čerenkov signal on the vertical axis and the total scintillation signal on the horizontal axis. We call this the "banana" plot. Note that $S > C$ always, since all charged particles contribute to S, and only the relativistic subset contributes to C. Dual-readout effectively measures each shower twice, in two different ways.

It is useful to understand that an electromagnetic energy deposit in a shower will move the event diagonally along the $C - S$ 45° line, and that dE/dx from a recoil or spallation proton will move the event along the S axis only. Therefore, putting more free H into the calorimeter (e.g., as scintillator) will capture more elastic $np \to np$

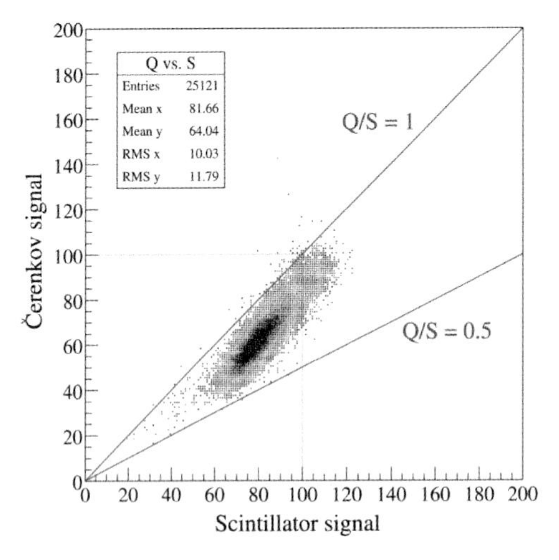

Figure 3.18 Dual-readout fiber signals for 100-GeV π in the DREAM module. The Čerenkov signal (C) is largely the electromagnetic part, and the scintillation (S) is all charged particles of the shower.

scatters and move the "banana" in Figure 3.18 into a more vertical position. When enough H is present, and you wait long enough, about 100 ns, for the neutrons to scatter down in energy, the banana is moved to a completely vertical position. This is called "compensation", a narrow scintillation signal that does not depend on the Čerenkov signal, *that is*, is independent of the hadronic shower fluctuations in π^0 vs. $\pi^\pm$ production. Similarly, leakage in the form of neutrons will move an event to the left, while electromagnetic leakage of showers at the periphery of the calorimeter will move the event diagonally down. Leakage in the small DREAM module is the essential reason that analysis of DREAM data, however excellent the analysis, is difficult and not completely unambiguous.

The (e/h) values, that is, the relative EM-to-hadronic responses for the separate Čerenkov and scintillation calorimeters, are determined from data [52] to be

$$\left(\frac{e}{h}\right)_C \equiv \eta_C \approx 5 \quad \text{and} \quad \left(\frac{e}{h}\right)_S \equiv \eta_S \approx 1.4\,, \tag{3.12}$$

and they are energy independent as far as we can determine. The Čerenkov response and the scintillation response to incoming hadronic energy, E, can be assumed to be linear in EM and non-EM parts and written for the Čerenkov calorimeter as

$$C = \left[f_{em} + \frac{1 - f_{em}}{\eta_C} \right] E \tag{3.13}$$

and for the scintillation calorimeter as

$$S = \left[f_{em} + \frac{1 - f_{em}}{\eta_S} \right] E\,, \tag{3.14}$$

where f_{em} is the fraction of the shower that is electromagnetic. For example, a hadronic shower with only $\pi^\pm$ and no $\pi^0 \to \gamma\gamma$ would have $f_{em} = 0$. For each shower, C and S are measured, and the response functions in Eqs. (3.13) and (3.14) are solved for f_{em} and E. The best estimate of the shower energy is

$$E = \frac{\eta_S E_S(\eta_C - 1) - \eta_C E_C(\eta_S - 1)}{\eta_C - \eta_S}\,. \tag{3.15}$$

A more compact way to write this is

$$E = \frac{S - \lambda C}{1 - \lambda} \quad \text{with } \lambda = \frac{1 - (h/e)_S}{1 - (h/e)_C} \sim 0.3 \text{ for the DREAM module.}$$

Data have been taken in the DREAM module with BGO crystals in front (similar to the 4th configuration) shown in Figure 7.4, and these results have been published [55]. A distribution of jet energies for 200-GeV π^+ is shown in Figure 3.19a and compared to a simulation of the 4th 4π-detector in Figure 3.19b. There are many differences between these data and the simulation, which had their own different purposes, but the combination of two very different dual-readout calorimeters

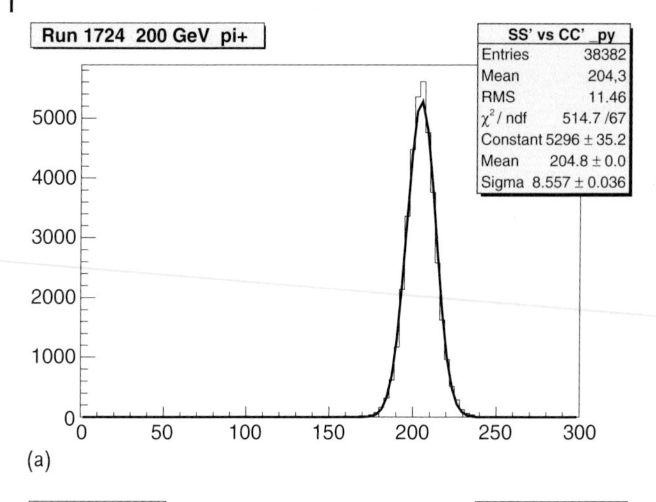

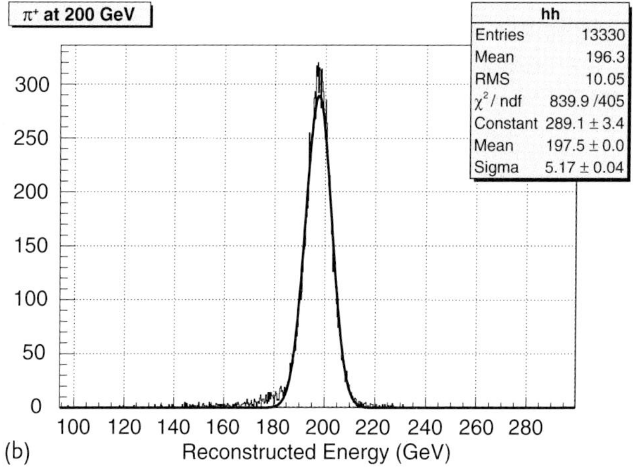

Figure 3.19 Measured and simulated energy resolution of dual-readout calorimeters. (a) The results of CERN beam test (Figure 7.4) exposed to 200-GeV π^+, and (b) simulation of a similar configuration of BGO plus deep fiber calorimeter. There are several differences. In (a), the DREAM module suffered neutron leakage laterally, which was partially compensated by plastic scintillator counters surrounding the module. In (b), the calorimeter system was a complete 4π detector without leakage and with a time-history readout of the plastic scintillating fibers and, therefore, a measurement of the neutron content shower by shower. These two response plots are, taking these several differences into account, consistent with each other.

(crystal and fiber) into a single calorimeter system that functions with the resolutions given by either one of these (data or simulation) is a statement about the robustness of dual-readout calorimeters. It is also evident that both the data and the simulation are Gaussian in their response, and that the resolution is very good, about 4% in data (limited by leakage from the small DREAM module) and about 2.6% in the simulation.

Particle Flow Analysis (PFA) Calorimetry

The idea of particle flow calorimetry was proposed by Morgunov [58] and Videau [59] as a new paradigm in calorimetry: use the combined momentum information from the tracking system (for momentum measurement of $\pi^{\pm}$, $K^{\pm}$, p, and $e^{\pm}$), from the electromagnetic calorimeter (for energy measurement of γ), and from the hadron calorimeter (for energy measurement of n, K_L^0). These three systems are spatially disjoint, and for Pb or W calorimeters, the depth development of γs and hadrons differ by a factor of 30, as seen in Figure 3.20. The problem reduces to one of combining this information into, for example, the energy of one jet. Since the showers from the neutral particles will overlap within the calorimeter volume with the showers from the charged tracks (whose energy has already been counted in the momentum measurement), the calorimeter geometry must have unprecedented granularity, approaching the cm^3 scale per channel, in order to accomplish this in software reconstruction. Early estimates of the expected jet energy resolution combined the momentum and energy resolutions of these three classes of particles in quadrature, weighted by the fraction of each class in jets. These fractions are approx. 62% charged tracks, 27% photons, and 10% neutral hadrons and about 1% neutrinos, with corresponding measurement resolutions of $\sigma_p/p \approx 10^{-4}\,p(\text{GeV/c})$, $(\sigma_\gamma/E) \approx 20\%/\sqrt{E}$, and $(\sigma_{h^0}/E) \approx 70\%/\sqrt{E}$. This thinking leads to an expected jet energy resolution expressed as

$$\sigma_{E_{\text{jet}}} \approx \sqrt{(0.62)^2 \times \sigma_p^2 + (0.27)^2 \times \sigma_{E_\gamma}^2 + (0.10)^2 \times \sigma_{h^0}^2}\,,$$

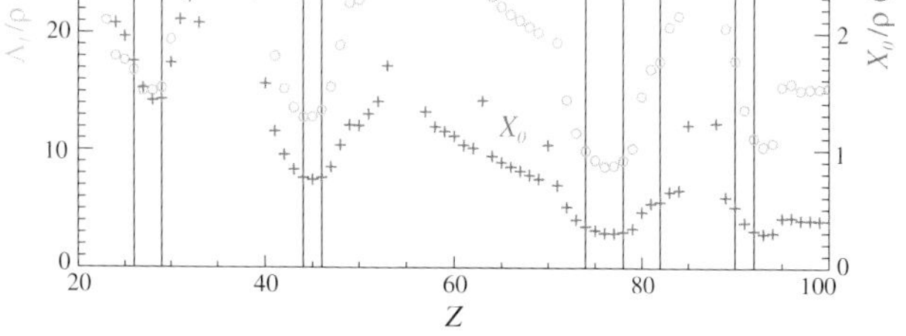

Figure 3.20 The radiation lengths (X_0/ρ, in centimeters, right-hand scale, symbol [+]) and the nuclear interaction lengths (Λ_I/ρ, in centimeters, left-hand scale, symbol ∘) of all atoms $20 < Z < 100$. For calorimeter purposes, the interesting elements are at the minima; note that for hadronic calorimeters, Cu is superior to Pb, while the reverse is true for an electromagnetic calorimeter. This plot is an interesting reflection of the Pauli Principle for both the electrons of atoms and the nucleons of nuclei. Plot from D. Groom, PDG.

resulting in an overall jet energy resolution of approx. $19\%/\sqrt{E}$. This procedure is misleading since the numbers 0.62,0.27, and 0.10 themselves fluctuate, sometimes by large amounts. The software reconstruction of jets and the untangling of energy deposits in the volume of the calorimeter is a great challenge, and an additional term is added to the above formula, $\sigma_{\text{confusion}}$, to account for the imperfect capability to properly apportion all energy deposits. This confusion term is a function of jet energy, channel volume sizes, and algorithms. Active R&D is being pursued on PFA calorimeters of several kinds [60, 61] for detectors at CLIC and ILC. Current results are shown in Figure 3.21a for a hadronic energy resolution (in terms of rms90) of $60\%/\sqrt{E}$ stochastic term and a 2.5% constant term.

However, it must be pointed out that nothing like a resolution of $19\%/\sqrt{E}$ has been achieved, even in a simulation with a perfect calibration [60], and the resolutions quoted in these analyses are not Gaussian resolutions but values of so-called rms90. Arguments have been advanced [60] that beam tests of PFA calorimeters are not necessary since the hadronic Monte Carlo simulation codes can be "validated" in a beam test with single pions and the performance of the whole 4π detector confidently simulated. This would be a severe – and dangerous – departure from common practice in experimental high energy physics. The complexity of a calorimeter cannot be fully simulated (some would say not even correctly simulated), and a proper understanding of calorimeters depends predominantly on the behavior of low energy particles in complex media. Every experiment to date has been tested before approval.

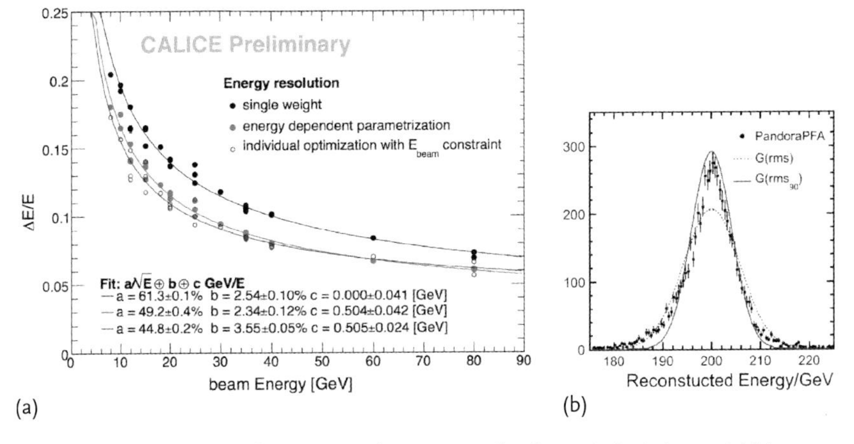

Figure 3.21 The available information on the energy resolution of PFA calorimeters. The upper curve, representing the data, uses a weighting procedure similar to that used by Abramowicz [71] and Milstene [63] and yields a stochastic term of 62.4% and a constant term of 2.5%. The other curves use procedures that cannot be used in a jet environment. (b) The vertical axis in panel (a) is not a Gaussian fitted rms but a quantity called rms90, defined as the rms of the smallest 90% of the events, that is, 10% of the events in the tails are excluded from the rms calculation. The relation of rms90 to a Gaussian σ and to the distribution as a whole is illustrated in (b). See also [72].

3.3.2.1 Digital Calorimeters

Digital calorimeters have been tried, and abandoned, for good reasons.
– Richard Wigmans

Digital calorimeters being considered for future colliders are largely driven by the particle-flow designs in which a large number of channels as small as $1\,\text{cm}^3$ are required for shower discrimination and association with the charged tracks measured in the tracking system. Given this prior need, digital calorimeters are a good idea since the electronics and readout are simpler if only a signal above threshold is required, as opposed to a full digitization of an analog signal. Single-digit 1-bit readouts (0–1, one threshold) have been enlarged to include 2-bit readouts (0–3, two thresholds), which effectively combine the benefits of both analog and digital signal acquisition.

For an analog readout, both the analog-zero-signal-level pedestal must be known and the gain must be known, in whatever convenient units one chooses, for example, mV/*mip*. A digital calorimeter replaces knowledge of the gain by knowledge of a threshold level, which itself will correspond, for example, to some fraction of a *mip*. This is not necessarily easier or more trustworthy, although it can be argued that the electronics is simpler.

For two or more *mips* in the same channel, the output is the same 1 bit, so the output signal from a whole shower will saturate as the energy and the number of shower particles increases, that is, the total signal in digits, N, increases more slowly than E, *for example*, $N \approx E^{\alpha}$, where $\alpha < 1$ and α depends critically on the volume of the channels relative to the effective volumes of electromagnetic showers. This track density is widely different for hadronic and electromagnetic showers, and different still for the spatial distributions of spallation protons and MeV-energy neutrons from nuclear breakup.

In most hadronic calorimeters, the instrumental response to electromagnetic showers is larger than the average response to hadronic showers, $e/h > 1$, and one way to attain "compensation", or $e/h = 1$, is to suppress the electromagnetic response by some means. A digital calorimeter achieves this quite easily since small electromagnetic showers from $\pi^0 \to \gamma\gamma$ decay within the hadronic shower, resulting in spatially dense cylinders of $e^{\pm}$ tracks, and only 1 digit is recorded per cm^3. In fact, this means of compensation was demonstrated in the WA1 calorimeter by Abramowicz [62] in software by simply reweighting channels with large analog signals, since these were channels with preferentially electromagnetic energy deposits; as discussed by Wigmans [29, Section 4.9], both the energy resolution and the linearity of the calorimeter were substantially improved. The same result was obtained by Milstene [63].

Quantitatively, if the average saturation results in a measured digital signal $N \approx E^{\alpha}$, then the energy resolution is quickly found from the derivative of $\ln N = \alpha \ln E$, or $dN/N = \alpha(dE/E)$, or

$$\frac{\sigma_E}{E} = \frac{1}{\alpha}\frac{\sigma_N}{N},$$

so that the energy resolution is worse than the measured signal resolution, σ_N/N, by a factor of $1/\alpha$. It should be noted that α, whatever its value may be, is not necessarily a constant over the energy range of interest but may grow larger at high energies due to the narrowing of high energy subshowers and that α, which is an average over a whole shower for some number of physical channels, will be affected by fluctuations in shower development, especially fluctuations to higher local particle densities. This can be represented as an energy-dependent power, $\alpha(E)$.

Absorber materials Hadronic calorimeters are critically dependent on the nuclei of the absorber and only weakly dependent on the electrons and the nuclear charge through signal generation (Section 3.1.1). For a compact calorimeter, one would choose an absorber from one of the minima in Λ_I (circles) in Figure 3.20, but talk to a metallurgist first. Fe and Cu (or brass) are easy to handle and form; Ru and Pd are rare and too expensive; W is excellent, but expensive and very hard; Au is too expensive; Pb is inexpensive and easy to form; and Th and U are interesting, but carry a burden of environmental radiation hazard monitoring.

For hadronic calorimetry, it is interesting to note that copper (Cu) is an excellent absorber, better than lead (Pb), even though it is less dense, as seen in the minima of Figure 3.20. Practical concerns in the construction of a large-volume calorimeter are the hardness (Fe, W, U) or the softness (Pb) of the material, although Pb can be stiffened. For a "traditional" calorimeter of alternating plates of absorber and sensor volumes, these concerns can be easily addressed. For calorimeters now being considered for future colliders, *for example,* particle-flow calorimeters with cubic-centimeter volumes, dual-readout calorimeters with millimeter-scale fibers or multicentimeter-scale crystals, the small-scale fabrication required will necessitate attention to Young's modulus, machining and forming techniques, and resulting costs.

Importance of neutrons Again, for hadronic calorimetry, achieving energy resolutions at the 1% level will require measurement of the binding energy loss fluctuations as energetic hadrons break up nuclei. Both the Poisson fluctuations in the number of broken-up nuclei and the efficiency of measuring the resulting *debris* of spallation protons ($T \approx 100$ MeV) and the liberated neutrons ($T \approx 2-6$ MeV) lead to important and critical choices in the design of a calorimeter. It is good to avoid U since in a calorimeter half of its neutrons come from fission and are, therefore, uncorrelated with the energetics of hadrons in showers.

Dagwood calorimeters More layers in a calorimeter, Figure 3.22, is not necessarily better, and the frequent argument that more information can only improve calorimetry is seldom true. A detailed discussion of some common problems is given by Wigmans [29, Section 6.2] and also discussed in [48]. For calibration, a common scheme is to perform a least-squares fit to a set of coefficients (one per depth section) with either one of two χ^2 requirements: (*i*) minimize the rms deviations of the calorimeter signal from the beam energy or (ii) minimize the energy

Figure 3.22 More layers may be better for a "Dagwood sandwich", but not for a calorimeter.

resolution of the ensemble. These two choices in general give different calibration coefficients.

Even for simple electromagnetic calorimeters, the instrumental response changes with the depth, or "age", of the shower since, physically, the shower consists of high energy $e^{\pm}$*mips* in its early development and particle multiplication stage, but consists largely of MeV γs that Compton scatter in the later stages of the shower in deeper depth sections. The instrumental response can differ by 30% for these two very physically different ensembles of shower particles. Setting all calibration coefficients to unity is as good, or better, a choice as any other, since at least the totality of a shower will be summed and the relative fraction of MeV-γs to $e^{\pm}$*mips* will be relatively constant from shower to shower. Finally, when electronic noise is a contribution to the energy resolution, summing N depth sections increases the noise by $\sqrt{N}$.

Material distribution in tracking-calorimeter systems The material in a tracking system affects the momentum resolution by introducing a multiple scattering fluctuation into the sagitta through the curvature term

$$\delta k_{\mathrm{ms}} \approx \frac{2}{\sqrt{3}} \frac{\theta_{\mathrm{rms}}}{\ell} ,$$

but the material near the end of the track, near the entrance to the calorimeter, affects the tracking less and the calorimeter more through unmeasured energy losses before reaching the calorimeter absorbing and energy measuring volumes. It is in exactly this radial region, between the tracking and the calorimeter, where mechanical supports are required for tracking chamber alignments and support and, more importantly, for support of the massive calorimeter. Some experiments have found it necessary to insert preshower chambers to measure exactly these showers that begin in the tracking system and the support materials before the calorimeter, and this may be a necessary price. However, I argue in Chapter 7 that reducing the mass of the tracking system solves many problems simultaneously, including the need to build and insert a preshower detector, Section 4.1.2.

Jet reconstruction The high-precision reconstruction of jet four-vectors depends on all detector systems and is one of the hardest problems for a big detector. Since jets are produced by the fragmentation of a quark or gluon, a jet consists of a broad momentum spectrum of all particle types, including the heavy b and c quarks that are likely to yield $e^{\pm}$ and $\mu^{\pm}$ in subsequent decays of their B and D mesons. Tagging B or D meson decays at the vertex gives a clue that a lepton may be present in an event, including a $\tau^{\pm}$ lepton, which will result in one or more missing neutrinos along the flight path of the jet. Not only must the low momentum particles that curl up the tracking chamber be properly ascribed to their correct jet, but the $\mu^{\pm}$ that appear in the muon system must be properly associated. Given these several effects at once, it is clear that high precision measurements of track momenta and calorimeter energies are important for jet four-momentum measurements. In addition, precision jets are necessary to tag possible missing neutrinos in a jet. Lacking good energy measurements means that a real neutrino will be taken for an energy resolution fluctuation.

Finally, any big detector with both tracking and calorimetry will depend on tracking ($\mu^{\pm}$), calorimetry (γ, n, K_2^0), or both ($e^{\pm}$, $\pi^{\pm}$, $K^{\pm}$, p) in the measurements of individual particles. The measurement of jets depends on everything at once. The relative precision of tracking and calorimeter systems is shown in Figure 3.23. The reconstruction of jets is an old art, different for hadron and electron machines due to widely different background conditions, and always difficult. Recent work in the context of the 4th detector with dual-readout calorimetry by Mazzacane [64] has resulted in a jet–jet mass resolution of

$$\frac{\sigma_{M_{jj}}}{M_{jj}} \approx \frac{34\%}{\sqrt{M_{jj}}} ,$$

shown in Figure 3.24. This excellent mass resolution is largely due to the excellent hadronic energy resolution of dual-readout calorimeters and is responsible for the successful $W \to q\bar{q}$ separation from $Z \to q\bar{q}$ by raw mass resolution, shown in Figure 4.8. At this precision of approx. 1% at $M_{jj} \approx 1$ TeV, many other effects enter the problem, for example: (i) the magnetic field bends charged particles out of the jet cone, (ii) $\mu^{\pm}$ from primary decays (B, D mesons) or from $K^{\pm}$ and $\pi^{\pm}$ decays in the tracking system or the calorimeter penetrate the calorimeter and exit the detector, and (iii) $\pi^0 \to \gamma\gamma$ decays leave bits of unassociated electromagnetic energy in the calorimeter. All of these must be gathered together into the proper jets. For a high-precision experiment, the correlations in energy and angle among the jets of an event must also be estimated from the shared energies in common channels and from misappropriations of particle energy in the jet-finding procedure.

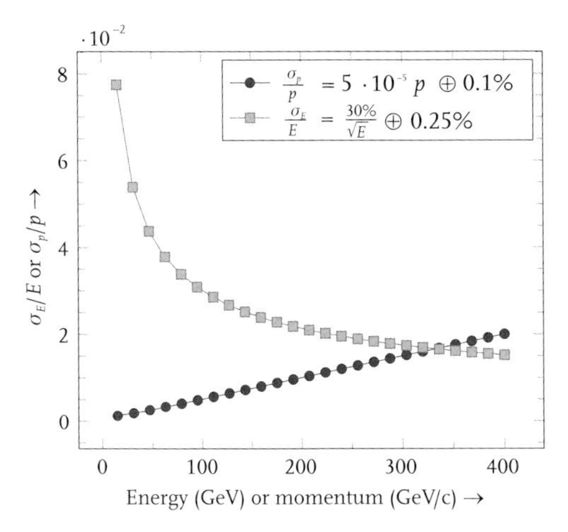

Figure 3.23 Comparison of tracking momentum resolution $\sigma_p/p \approx 5 \cdot 10^{-5} p \oplus 0.1\%$ to calorimeter energy resolution $\sigma_E/E \approx 30\%/\sqrt{E} \oplus 0.25\%$ for the best expected resolution attainable at a future collider. The crossover point is about 300 GeV.

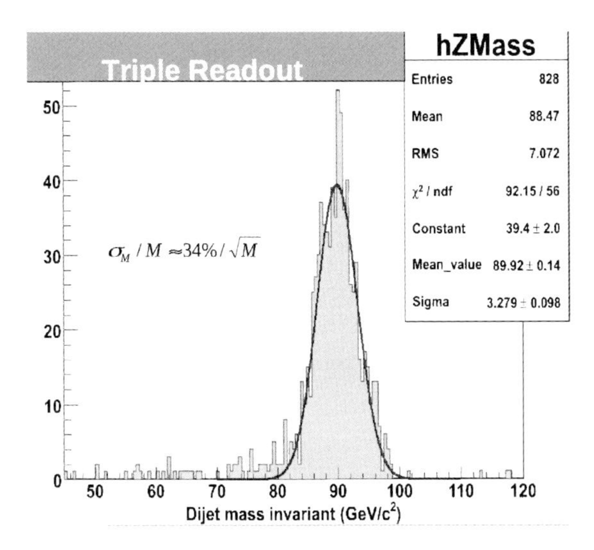

Figure 3.24 Mass resolution of two jets in dual-readout calorimeters and tracking system of the 4th detector. The header "triple readout" refers to the calorimeter reconstruction that uses the time history of the scintillating fibers, $S_{pe}(t)$, to estimate the neutron content of a jet and, in this way, to improve the energy resolution with this effective measurement of the binding energy losses incurred in the hadronic cascade. From A. Mazzacane [64].

3.3.3 Calorimeter Constant Terms

The usual way to represent the energy resolution of a calorimeter is the sum of a "stochastic" term (k) that scales as $1/\sqrt{E}$ and is governed by the statistics of particle counting, and a "constant" term (c) that is independent of energy:

$$\frac{\sigma_E}{E} = \frac{k}{\sqrt{E}} \oplus c \,.$$

The stochastic term is usually easy to calculate and understand, especially for electromagnetic calorimeters, but the constant term depends on many small things, each adding up to a total that must be less than 1% (for a good calorimeter). For high energy colliders this constant term will become vitally important as the stochastic term becomes small. For example, for $k \approx 40\%$ at a jet energy of 500 GeV, the stochastic term is 2%, and any hope of reaching a calorimeter resolution of 1% is completely dependent on the constant term being less than 1%, in fact, at least as small as $c \approx 0.5\%$. A calorimeter constant term can have many sources, and only the three obvious ones are discussed here.

Due to volume nonuniformity The zeroth rule of any calorimeter is that the volume uniformity must be constant. A 2% deviation from uniformity is a 2% constant term. This nonuniformity can arise from manufacturing mistakes, inactive regions around the edges of sensor chips, attenuation of light in a scintillating tile, crystal, or fiber, variations in the refraction of light on crystal surfaces, nonuniformity of response across the face of a photomultiplier tube (PMT) photocathode, calibration mistakes, *in other words*, almost everything can get you. The challenge is extreme in that a 1% constant term is already too large for a future precision detector.

Volume uniformity of calorimeter response is absolutely critical.

Due to noncompensation Any understanding of the constant term, or the nonconstancy of the constant term, requires a deep understanding, and I recommend the concise paper by Wigmans [48]. In addition to the physical properties of the calorimeter, a noncompensating calorimeter also has a contribution to the constant term proportional to the deviation of (e/h) from one, sometimes written [49, Equation 28.32] as

$$c \approx |1 - (h/e)| \times \mathcal{F}(f_{\mathrm{em}}) \,.$$

This "constant term" depends on energy since the average electromagnetic fraction f_{em} depends on shower energy, here denoted by the function $\mathcal{F}(f_{\mathrm{em}})$, which can be written as a power law in shower energy. This is a complex discussion best left to experts. For purposes of a good calorimeter inside a good detector, it is best to have a dual-readout calorimeter, or at least a compensating calorimeter to avoid, at the outset, all of these complex problems. It is also best to be on the safe side since, although a physics understanding of these complex issues may be available and even correct, the construction of a big calorimeter will involve choices and decisions that may compromise volume uniformity or the ability to achieve effective compensation, even in small ways, and small ways are important because 1% is a small number.

Compensating calorimeters solve many problems at once.

Due to leakage, or noncontainment There are many plots of the leakage of particle energies out the back of calorimeters as a function of the particle energy and the depth of the calorimeter, and lines drawn for 10 or 1% average leakage. There is nothing wrong with these, but they are misleading in the sense that the average is not important. In fact, if the fractional leakage were constant for every event, it would not matter at all for the energy resolution, no matter how large the leakage. It is the *fluctuations in leakage* that are important, and a rule of thumb is that

> the rms fluctuations in energy leakage are approximately equal to the average leakage

or, in other words, the fluctuations in the leakage itself are about 100%. Hence, a mean leakage of 2% incurs a constant term of 2%.

Success stories Two of the best hadronic beam test calorimeter modules ever built were the original compensating SPACAL module (Figure 3.17) and a test module advocated by the GLD concept detector [65] for the ILC, which achieved, see Figure 3.25, constant terms of $c \approx 0.8 \pm 0.3\%$ and $c \approx 0.9 \pm 0.9\%$ for electromagnetic (e^-) and hadronic (π^-) showers, respectively.[30] The stochastic terms are $(23.9 \pm 0.3)\%/\sqrt{E}$ and $(46.7 \pm 0.6\%)/\sqrt{E}$ for electromagnetic and hadronic showers, respectively. These numbers are completely consistent with the earlier SPACAL results [66]. Extending the excellent energy resolution of the GLD calorimeter to a PFA analysis of simulated events at the Z^0 resulted in a total energy, or mass, resolution at the Z^0 of $38\%/\sqrt{E}$, see [65]. This experimental beam test result is tantalizingly close to the physics objectives of a future collider calorimeter, which is $30\%/\sqrt{E}$ plus a small constant term.

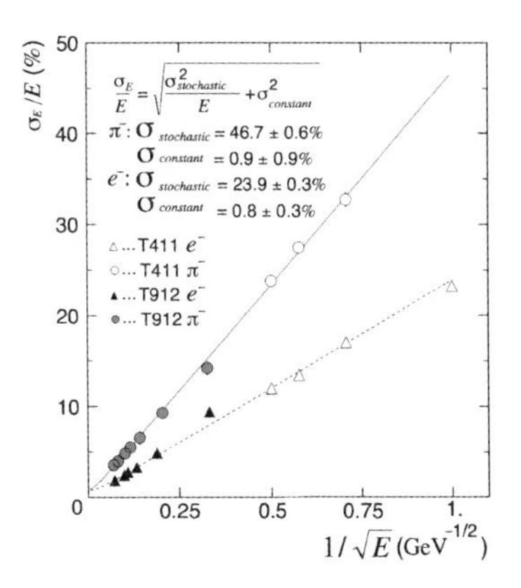

Figure 3.25 The stochastic and constant terms from the GLD calorimeter test of a compensating Pb-scintillator calorimeter [65].

30) It must be noted that the GLD calorimeter was carved into small volumes for double use as a PFA calorimeter, and it is possible that these constant terms of order 1% had small contributions from the nonuniformity among these many small volumes.

There is usually a correlation between a large constant term and large systematic uncertainties in the hadronic, or jet, energy scale in a physics experiment, and some experiments have physics results ultimately limited by this systematic uncertainty. All of the efforts and expense put into higher energies and higher luminosities at a collider will be defeated by a constant term in the calorimeter.

3.4 Time, or Velocity, Measurements

The best time standard is the velocity of light, c, and there are two ends: the production end and the detection end. At production, scintillation light is a quantum process with a mean exponential lifetime (Section 3.1.1), and the fastest scintillators are in the subnanosecond region. Čerenkov light is also a quantum process, but its time development is instantaneous on the time scale of any detector and can be taken as zero.

At the detection end, the neutral photon must interact before it can be detected, and the best means is the photoelectric effect in the photocathode of many possible and ingenious devices: PMTs, silicon photomultipliers (SiPMs, also known as multipixel photon counter, MPPC), microchannel plates, metal channel plates, and several types of photodiodes. In a typical PMT, the photoelectron falls through potential differences between multiple dynodes, traversing centimeters of path, and the total time between the generation of the photoelectron and the signal is the "transit time", typically 50 ns. The rms spread is the "transit time spread", which can be 2 to 3 ns. Shrinking the dimensions reduces these times, including the rms spread. The advantage of SiPMs and microchannel plates is that they are physically small objects, and the rms time resolution is very small, approaching 10 ps. The best references are the periodic conferences on photon detection, PD09 [67].

Direct clock The detection of optical light at velocity c is the leading method for measurement of the arrival time of a particle at a detector, where the velocity of light is 1 foot per nanosecond, or 30 cm/ns, and so the time for a particle to traverse a large detector is typically 20 ns. A physical particle will take a time T(ns) to arrive at the detector, and this time depends on the particle mass (Section 4.9 on time-of-flight mass measurement), its particular geometrical trajectory (a helical trajectory, for example), whether it came from the decay of a long-lived slowly moving particle (e.g., the photons from $\Lambda^0 \to n\pi^0 \to n\gamma\gamma$ decay), or $\pi^\pm$ and γ from the decays of K^0s in which the total flight path of the parent plus daughter is not a straight line from the origin. The particle must generate the optical light to be detected, and there are two main possibilities: Čerenkov light that is generated instantaneously (compared to any detector) in a transparent optical medium (glass, plastic, gas, water, oils, etc.) and scintillation light that is generated in the deexcitation of certain molecules and atoms that have been excited by the dE/dx mechanisms as the particle loses energy to these atoms. The best time-of-flight measurements to date are in the 10-ps range,[31] and plans exist to both improve this time and build fast photocathodes with large m^2 areas [68, 69].

Čerenkov light is the optical analog of a sonic boom in sound waves, and the quantum time is small compared to any photo-converter so that Čerenkov light is effectively instantaneous for any detector.

Scintillation is complicated by the initial excitation process followed by quantum deexcitation, which must necessarily have a mean lifetime, τ. For time-of-flight measurements one wants a fast scintillator, or small τ, and good plastic scintillators have lifetimes $\tau < 0.5$ ns. Of course, there are many much slower scintillators such as NaI(Tl) with $\tau \approx 230$ ns and BGO with $\tau \approx 300$ ns. The choice of scintillator depends on many factors, which we only list here: lifetime τ, luminosity (photons/MeV of dE/dx), density, radiation hardness, transparency to the scintillation light, cost, hydrogen (free proton) content for neutron sensitivity, wavelength sensitivity, and so on. The gain and quantum efficiency of the photoconverter must be matched to the scintillator. These issues and more are discussed in a recent conference paper [68, 69].

There are several optical photoconverters that maintain excellent time responses, that is, deliver an electronic signal promptly when a photon ejects an electron from the photocathode; see a recent conference paper PD09 [67]. Many important properties of scintillators are listed in the PDG [49, Sections 28.3 and 28.4].

For versatility, the PMT, invented in 1934, is excellent in almost all respects: versatile geometries and sizes, very low noise, robust, and simple. The only serious weakness of a PMT is that it cannot be used in $\boldsymbol{B}$ fields above a few hundred Gauss and only up to 0.1 T with "mu metal", or soft highly permeable iron, shielding.

There is a huge interest in and substantial usage of "silicon PMs" (SiPM), in particular in the CALICE calorimeters [61]. However, SiPMs are not trouble free. Depending upon the electronic circuitry in the read-out, there can be a high "dark rate" ("dark" means detector signals when no light is incident) and after-pulses (secondary pulses following the main pulse in time by tens of nanoseconds). On the practical side, the effective photocathode area is usually small, $(3\,\text{mm})^2$, and the cost is about \$50 per mm^2 (in 2009 units), but falling. There are many photoconverters under active R&D, such as microchannel plates, metal channels, and new SiPMs. The situation is so fast and fluid that the only sensible course is to follow work shown at current conferences such as the photodetector (PD) series, the latest being PD09 [67].

In big detectors at high beam rate colliders, "time stamping" the signals from a detector is important so that particles from one collision are not confused with particles from a previous or later collision, and this is particularly important at the LHC with a beam crossing time of 25 ns. Future colliders will offer a wide range of crossing times: the CLIC e^+e^- collider will cross bunches every 0.4 ns, in other words, essentially a continuous interacting beam as far as the detector is concerned; the ILC e^+e^- collider will cross every 337 ns, comfortably within the

31) For example, [70].

speed and resolving time of most modern detector technologies; and the Muon $\mu^+\mu^-$ Collider will cross every 6 μs, a very comfortable rate for detectors.

In calorimeters that depend on compensation by neutrons, the PD must be able to see single recoil protons from $np \rightarrow np$ scatters long after the event is finished, and SiPMs are a difficult choice with their dark rates and after-pulses occurring in the tens-of-nanoseconds range.

Čerenkov angle as a velocity measurement Čerenkov light is emitted by any charged particle with velocity $\beta = v/c$ greater than $1/n$ in an optical medium, with zero light emission and zero Čerenkov angle at $\beta = 1/n$. Both the intensity and Čerenkov angle increase as β exceeds $1/n$. Specifically [49, Section 27.7], the Čerenkov angle is

$$\theta_{\mathrm{C}} = \cos^{-1}\left(\frac{1}{n\beta}\right) ,$$

and the light intensity is

$$\frac{d^2 N}{dx dE} = \frac{\alpha}{\hbar c} \sin^2 \theta_{\mathrm{C}} \approx 370 \sin^2 \theta_{\mathrm{C}} \left(\mathrm{eV}^{-1}\mathrm{cm}^{-1}\right) .$$

Since both the angle and the intensity depend on β, the Čerenkov emission is a measurement of velocity, but only usefully sensitive near threshold. Čerenkov light is used very successfully at $B\bar{B}$ factories such as BABAR, *for example*, DIRC, where particle identification of $K^\pm$, in particular, is critical and the particle density backgrounds are low. These devices have not been proposed (yet) for future colliders where the track backgrounds from jets are huge and it is sometimes argued that low momentum particles are not important since the physics is at high energy. This is a wrong argument since, at the very least, the known and expected production of top, $t\bar{t}$, will end up as $K^\pm$ mesons through the decay chain $t \rightarrow b \rightarrow c \rightarrow s \rightarrow K$. In addition, the b quark will result from heavy speculative particles and again result in Ks in the detector. These Ks will be low momentum in the GeV range and ripe for identification by dE/dx and time-of-flight techniques.

Much more useful is the fact that Čerenkov light is generated instantaneously (on the time scale of any detector) and is very useful as a "zero-time" signal. In addition, in optical media such as plastic or glass ($n \approx 1.5$) or in H_2O ($n \approx 1.33$), Čerenkov light is only generated by relativistic particles such as electrons, $e^\pm$, and in a hadronic shower consisting of slow spallation protons, moderate energy $\pi^\pm$, p, and slow $K^\pm$, a Čerenkov signal will distinguish this purely electromagnetic component for the hadronic components. This is highly useful in dual-readout calorimeters, Chapter 7.

3.5 Signal Distribution among Channels

This is a small point about measurements, but every measurement is so many pC or mV in an electronics channel, which corresponds to a silicon strip or pixel, or to a calorimeter volume, or to the light in fibers.

Spatial Resolution Depends on the Signal Distribution among the Measured Channels

All measurements end up as signals in bins: energies in calorimeter channels, pulse heights in silicon strips, pulse heights in the time bins of a time-to-digital converter (TDC), or the channels of a fast analog-to-digital converter (flash ADC). The resolution in the centroid of these channels depends on the distribution of the signal among the channels, as illustrated in the figure. If the signal is all in one channel, the best estimate of the centroid is the exact center of the channel ($x = 0$ in this case), and the variance (the square of the rms expected uncertainty) is the integral of the response function, $f(x)$, times the square of the deviation of x from the mean, normalized by the integral of $f(x)$,

$$\sigma_x^2 = \int_{-w/2}^{+w/2} f(x)|x-0|^2 dx = \frac{1}{w}\int_{-w/2}^{+w/2} x^2 dx = \frac{w^2}{12},$$

so that $\sigma_x = w/\sqrt{12} \approx 0.3\,w$.

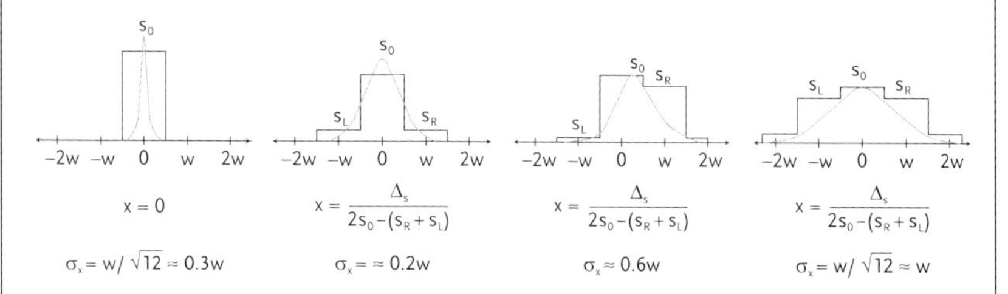

For signals distributed over several channels, a simple parabolic fit yields an estimate of the centroid as

$$x \approx \frac{x_R - x_L}{2x_0 - x_R - x_L}$$

and centroid uncertainties are shown for illustration in the figure (assuming $\sigma_0/s_0 \sim 10\%$). The best resolution is obtained for side channels that contribute information on the centroid. Of course, this exercise is not a substitute for a least-squares fit using the proper response function, $f(x)$, of the instrument.

The physical signal is a deposition of ionization electrons in a medium that is drifted through silicon and amplified, or the generation of photons in an optical volume that are transported and amplified in a photoconverter, or a voltage pulse on a wire that is digitized. Almost always the physical signal (a track or a shower in a calorimeter) activates more than one channel, and almost always there are adjacent channels in space or time that represent the total signal of a track or a shower. The distribution of these signals among the channels is important for both signal-to-noise, pattern recognition, and resolution.

If the signal is concentrated in one channel, the signal-to-noise and pattern recognition are optimal, but the resolution on the centroid of the physical signal is limited to the channel width (in space or time) divided by $\sqrt{12}$. A physical signal that is spread among three channels has three times the noise and may be slightly deleterious to pattern recognition, but the resolution on the centroid can be better. The details are so particular that, aside from generalities, not much can be said. However, it is desirable in a detector to concentrate the physical signal into a small number of channels, and for calorimeters the concept of "projective geometry" is intended to achieve this, greatly improving pattern recognition and signal-to-noise but making the physical construction of a calorimeter more difficult. Some calorimeters have settled for "semiprojective" geometries in which small nonprojective volumes are summed along radii from the origin to mimic a projective geometry.

3.6 Problems

1. Estimate the measured signal from a 6-MeV neutron in a plastic scintillator. The most efficient mechanism for energy loss by a low energy neutron is by elastic scattering from free protons (in a H–C plastic scintillator). On average, a neutron loses a fraction

$$\frac{\Delta E_n}{E_n} = \frac{2A}{(A+1)^2} \to \frac{1}{2} \text{ for } A = 1 \, .$$

of its kinetic energy per scatter. The signal will be the emitted scintillation light, which is affected by Birks recombination. Chose Birks' constant $k_\text{B} \approx 0.01\,\text{g} \cdot \text{MeV/cm}^2$.

Edward Witten made a beautiful case for high resolution measurements at a linear collider (International Conference on Linear Colliders, Paris, 19–23 April 2004, p. 95) He did not specify numbers, however. The next two problems (Problem 2 for spatial measurements and Problem 3 for calorimeters) are open-ended questions about the ultimate resolutions allowed by atoms and quantum mechanics.

2. What are the fundamental limits on the measurement of the sagitta, σ_s, for four detector planes, one at $z = 0$, two at $z = 0.5\,\text{m}$, and the fourth at $z = 1\,\text{m}$, that is, $\ell = 1\,\text{m}$? Consider all atomic media at all possible densities and in all

possible manifestations (e.g., gas, liquid, solid, semiconductor, superconductor, etc.).

For example, at zero density, there are zero atoms and no measurement, $\sigma_x \approx \infty$. At very high densities, there is a large signal from the electrons of many atoms, but multiple scattering on the nuclei of those same atoms limits the spatial information from the electrons, and $\sigma_s \approx$ large.

a) Consider the case of a single-atom detector. You know where the atom is and you can detect (electronically or optically) the interaction of a particle with this single atom. What (Z,A) would you choose, and how would you make the measurement?

b) Is there a fundamental crossover between fluctuations in the ionizations of atoms and fluctuations in the multiple scatters?

c) Can you find a minimum for σ_s in the space of all atoms (A,Z), of all densities, in all combinations, and for all means of detection?

3. For a calorimeter measurement of energy, you want to count the total number of particles and their species (i.e., $e^{\pm}, \pi^{\pm}, p$, etc.). Suppose you can do this perfectly.

   a) What are the fundamental limitations on energy resolution for electromagnetic energy measurements? Consider electroproduction of hadrons, the cross-section minimum at 1 MeV for $\gamma - (Ze)$ interactions, and differences in the interactions of e^- and e^+ in atomic media.

   b) What are the fundamental limitations in the energy resolution for a hadronic particle? Consider unknowable nuclear effects, the production and escape of neutrinos from $\pi \to \mu\nu_\mu$ (and others), the fluctuations in particle species (π^0 vs. $\pi^{\pm}$, etc.), and any fundamental inefficiencies in the measurements of particles as they approach zero energy.

4. As a way of thinking, consider the (impractical?) calorimeter that consists of pure hydrogen, a single electron bound by 13.6 eV and a single proton (no binding energies, no neutrons, no α particles or spallation protons, etc.). Going to a liquid gets the density up to $\rho \sim 0.07\,\mathrm{g/cm^3}$. Design your means of particle measurements, and estimate the ultimate electromagnetic and hadronic energy resolutions.

5. If you are happy with the results of Problem 4, consider a molecule, say, H_2O or CH_4.

6. If you want to think big, consider the interstellar medium as a calorimeter of H atoms of density $\rho \sim 1$ H atom/m^3.

7. A good vertex chamber contributes to a big detector in four important ways: it measures the impact parameters of decay tracks; it provides a precision spatial measurement near the interaction point for tracks that do not come from the primary vertex; it makes a momentum measurement even though ℓ^2 is small; and it provides a well-defined beginning track to project out to, or be linked to, a track in the main tracking chamber. All of these are limited by multiple scattering in the vertex chamber and, therefore, all designs of future chambers will minimize the amount of material.

   For example, the SiD and ILD letters of intent specify $0.1\% x/X_0$ and $0.74\% x/X_0$, respectively, near $\theta \sim 90°$. These are aggressive numbers compared to Figure 6.13, and the spread in the numbers reflects assumptions about both services and cooling.

   For a five-layer vertex chamber with a total material budget of 0.75%, inner radius $r_1 \sim 1.5$ cm, outer radius $r_N \sim 8$ cm, and pixel size $w = 20$ μm, and for a $p = 10$ GeV/c track:

   a) numerically estimate the two constants ξ and ζ;
   b) estimate the impact parameter resolution, σ_b (ignoring the η term);
   c) estimate the momentum resolution both using and not using the interaction point (which at the ILC is $\sigma_y = 4$ nm, $\sigma_y = 400$ nm); and
   d) reestimate σ_b for a material budget ten times larger.

Harry van der Graaf has made two novel suggestions (Vienna Conference on Instrumentation, Vienna, 16 February 2010) for future tracking instrumentation: (1) a tracking chamber without wires and without gas and (2) optical delivery of power to pixel chips. He also addresses cabling and cooling as the main contributors to material budgets in tracking chambers. The next three problems concern these issues.

8. The wires, silicon strips or pixels, and micropatterned structures (GEMGrid, TwinGrid, etc.) in present high precison tracking chamber designs can be replaced by microchannel plates (MCPs) positioned just below a thin foil for "secondary electron emission" to catch the ejected electrons, all inside a vacuum. The single electron is amplified by the avalanche in the MCP. The secondary emission foils would be 100 nm of CsI, but the MCP would be more. Design a simple four-plane tracking system ($\ell = 1$ m), and estimate its sagitta resolution. Available parameters are the microchannel spacing (affects spatial resolution), channel depth (affects electron gain), and the SEM foil-to-MCP geometry (affects time resolution, which can be $\sigma_t \sim 0.2$ ns).

9. The main difficulty with silicon tracking systems is the delivery of power and the extraction of the resultant heat. R&D is ongoing for so-called "power pulsing" of silicon tracking systems in colliders with long beam-down times (e.g., the ILC has long 0.20-s periods without beam; the beam trains of 3000 bunches spaced by 337 ns arrive at 5 Hz.). At the ILC, the power would be turned off

during the down time, reducing the power load by about a factor of 100. Take one 20 $(\mu m)^2$ pixel with a depletion region of depth d:

a) For $B = 5\,T$, $d = 50\,\mu m$, and $i_{drain} = 10\,nA$, what is the Lorentz force on this one pixel?
b) For a ladder of 25 M pixels, what is the force if all the pixel forces are aligned?

10. Graaf has suggested optical powering of pixel chambers in which 0.1 W of IR optical power is delivered to photodiodes on a pixel chip. There are plenty of designs to think about here. Consider the simplest case: one fiber to one chip that services 10^6 pixels. Consider the quantum efficiency of the photodiodes, power conversion efficiency between the photodiode and the silicon circuit, and intrachip distribution. There must still be a depletion region requiring a voltage and with a drain current. Consider fiber volume and mass for a large tracking system.
11. To solve the other half of the heat load problem, Graaf suggests an optical data acquisition for a readout of detector information. This is too sophisticated to be put into a problem, but the main and most simplistic notion is that transport of both data and power by photons is "free" (because they are bosons), whereas the transition to electrons immediately incurs Joule-I^2R power losses due to material resistivity. Without actually designing (or understanding!) anything, estimate the minimum power consumption of a 10^6-pixel chip equipped with fiber power-in and fiber data-out.
12. Technical optimism. Actually built, tested, and calibrated detector systems always lack the promised performance of unbuilt, untested, and futuristic detectors. Look around at some of the big experiments and develop a notion for the factor between initial dreams and final detectors. You may have to dig deep to find the initial designs, usually in the pre-TDR and preproposal stages, of an experiment. You will likely be deeply impressed by how close to one this ratio usually is. In the design of an experiment, one has to carefully and honestly distinguish the varying and changing stages of the R&D levels on different detector systems, and not be misled by promises and simulations. In the end, there is never a substitute for a beam test with actual particles into an actual detector with an actual readout. This problem is open-ended, of course, and the ratios between dreams and reality will vary widely between tracking systems and calorimeters (usually worse than tracking) and will vary between slow-electron, fast-proton machines (proton machines are usually harder). The judgments one makes about detectors are critical for a big experiment, and Appendix A is an example of many detector ideas in 1960 to which we now know the answers: which ones were successful, and which ones where not.

4
Particle Identification

"We reiterate the superior particle identification capabilities of the dE/dx technique. In the nonrelativistic region, dE/dx depends on $1/\beta^2$ and so is superior to time-of-flight which depends on $1/\beta$. In the relativistic region, dE/dx can achieve sufficient resolution to identify π, K, and p for the full PEP energy region, even in jets where particle densities are high."

– Dave Nygren, TPC proposal to SLAC, 1976.

The valuable particles are the $e^{\pm}$ and $\mu^{\pm}$ for $W^{\pm}$ and Z^0 decays, which are generally isolated spatially from other event particles, isolated photons (γ) that could come from $H^0 \to \gamma\gamma$, and very high energy jets (from quarks and gluons). It is expected that putative exotic objects will also decay into Ws, Zs, γs, and jets. A typical interesting momentum is greater than half the W mass. The overwhelming background, even in a lepton collider without hadronic jets dominant in most events, are the charged pions, $\pi^{\pm}$, from quark and gluon fragmentation. All high momentum charged tracks look the same in a tracking detector, apart from small differences in ionization in the few-GeV region. It is my opinion that particle identification has received too little attention in big detectors up to now. It would be most desirable if *all partons* of the standard model could be identified with high efficiency and, simultaneously, with low misidentification probabilities ("fake rates") from other more numerous particles.

The CMS experiment is about 7 m in radius and is a comprehensive detector, a sector of which is schematically illustrated in Figure 4.1. The behavior of all basic particles ($\gamma, e^{\pm}, \pi^{\pm}$, and $\mu^{\pm}$) is shown as they pass through the detector, and all the different interactions are displayed here. Starting from the left side at the interaction point (IP), and starting with the particle at the top:

- a photon (γ) leaves the IP, leaving no trace in the tracking system, and depositing all its energy in the electromagnetic calorimeter. No evidence of it is seen in the hadronic calorimeter;
- a neutron hadron (either K_L^0 or a n) leaves the IP, leaves no trace in the tracking system, penetrates the electromagnetic calorimeter, and deposits all of its energy in the hadronic calorimeter;

Particle Physics Experiments at High Energy Colliders. John Hauptman

ISBN: 978-3-527-40825-2

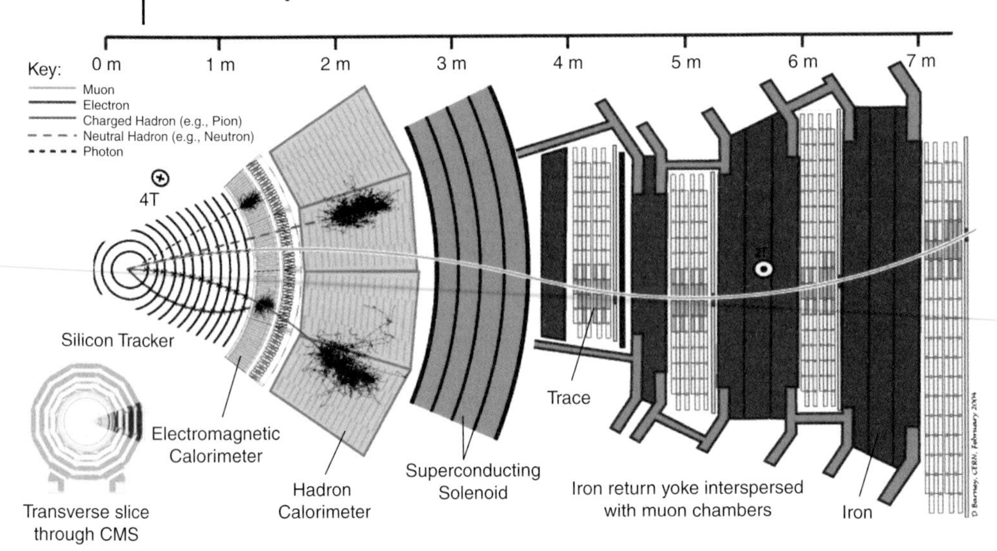

Figure 4.1 An illustration of the detector signatures of $e^{\pm}, \gamma, \pi^{\pm}, \mu^{\pm}$ in the CMS detector. From Efe Yazgan.

- a negatively charged muon, μ^-, leaves the IP, bends clockwise as $\boldsymbol{F} = e\boldsymbol{v} \times \boldsymbol{B}$ in the 4 T magnetic field, leaves traces in the tracking system along its trajectory, penetrates both the electromagnetic and hadronic calorimeters, penetrates the superconducting coil, penetrates the first layer of iron, leaves traces in the muon system chambers, bends counterclockwise in the magnetic field inside the saturated ($B_{\text{max}} = 1.8$ T) iron, leaves traces in all muon chambers, and leaves the detector volume;
- a negatively charged hadron (π^-, K^-, $\bar{p}$) leaves the IP, leaves traces in all tracking chamber layers, penetrates the electromagnetic calorimeter, and deposits all of its energy in the hadronic calorimeter; and
- a positron (e^+) leaves the IP, bends in the field, leaves traces in the tracking chambers along its trajectory, and deposits all of its energy in the electromagnetic calorimeter.

This is idealized, of course, but illustrates clearly how a big detector intends to identify the particles of the standard model so that, eventually, $W^{\pm}$s, Z^0s, and t quarks can be fished out of the data. The CMS detector takes data at an unforgiving machine, the Large Hadron Collider (LHC), at which bursts of particles leave the IP every 25 ns, and any interesting event, for example, with a μ^+ in it coming from a W^+, will have approximately 20 to 30 other "uninteresting" events (i.e., no $\mu^{\pm}$, no $e^{\pm}$, just pions from background jets) included in the read-out of the event. In addition to fast detector responses and 25-ns electronics to record interesting (or "triggered") events, particle identification and the capability to unravel particles will be highly important in all the LHC experiments (CMS, ATLAS, ALICE, and LHCb). The techniques shown here for CMS are conventional and known to work.

Problems arise when (*i*) hadronic particles interact in the electromagnetic calorimeter, (*ii*) photons convert to electron-positron pairs ($\gamma \rightarrow e^+e^-$) in the material of the tracking system, (*iii*) an electron undergoes bremsstrahlung in the material of the tracking system, (*iv*) a photon and pion are close to each other in the tracking system, mimicking an electron, (*v*) a charged pion undergoes charge exchange in the electromagnetic calorimeter ($\pi^- N \rightarrow \pi^0 N$) leaving almost all its energy in the calorimeter, mimicking an electron, (*vi*) a high momentum $\pi^0 \rightarrow \gamma\gamma$ decay leaves both photons very close to each other in the electromagnetic calorimeter, mimicking a single photon or, in general, a pile-up of tracks so that the event is confused. All of these problems (and more regarding calibration) are well known and their rates easily calculated.

In this section, we will discuss some nonconventional particle identification measurements that augment these more conventional ones and that have benefits for a high precision detector.

4.1
Discriminating Charged Leptons from Charged Pions

For future high energy colliders, efficiency and near-absolute identification of the leptons $e^\pm$ and $\mu^\pm$ is essential, simply because of the decays $W^\pm \rightarrow e^\pm\nu$, $W^\pm \rightarrow \mu^\pm\nu$, $Z^0 \rightarrow e^+e^-$, and $Z^0 \rightarrow \mu^+\mu^-$. And when the $W^\pm$ and Z^0 decay into $\tau^\pm$s, the $\tau^\pm$ subsequently decays into $e^\pm$ and $\mu^\pm$. In addition, there are speculated particles that decay into these three leptons, and almost every high-mass object of importance decays into $W^\pm$ and Z^0. This is critical at a hadron collider, such as the LHC, with a constant background rate of QCD jets that bathe the detector in pions, and therefore rejecting pions and keeping $e^\pm$ and $\mu^\pm$ is the critical first step for a detector. At an electron or muon machine, the QCD background comes from the physics events themselves, and is therefore far less critical, but not to be ignored. One's expectations for clean events and clean physics are much higher at an electron machine, and approximately the same standards for particle identification are sought. Some of the methods discussed here related to calorimeters are also discussed in more depth by Wigmans [29, Section 7.5]

4.1.1
Telling a $\mu^\pm$ from a $\pi^\pm$

The weakly interacting $\mu^\pm$ at high energy penetrates meters of iron with an occasional electromagnetic bremsstrahlung or pair-production event with the probability shown in Figure 3.4. In contrast, hadrons, in particular plentiful $\pi^\pm$ from jets in the event, are attenuated by about a factor of two for each nuclear interaction length, about $\lambda_{\text{int}} \approx 17$ cm in iron. This method is almost universally used to identify $\mu^\pm$ and the latest such detector with an iron absorber is CMS [73]. An alternative method, without iron, is the toroidal magnetic field of the ATLAS detector [74] or the dual solenoid of the 4th concept detector [75].

Muon identification and measurement in an iron (Fe) absorber In most big detectors, the magnetic field in the tracking-field volume is established by a solenoid, and the flux generated in this solenoid must be returned to the other end. An iron yoke is almost always used as the high-permeability flux channel, and this iron volume is simultaneously useful for the mechanical structure of the whole detector and, finally, as a charged pion ($\pi^{\pm}$) absorber that allows, mostly, only muons ($\mu^{\pm}$) to penetrate the 2 to 3 m of iron. The efficacy of this filter depends somewhat on the energies of the pions, but is mostly limited by the "punch-through" of actual hadronic pions, which can make repeated elastic or diffractive scatters in the iron. It should be noted that pions develop hadronic showers consisting partly of many more lower-energy pions and kaons that decay into muons, $\pi^{\pm} \to \mu^{\pm}\nu_{\mu}$ and $K^{\pm} \to \mu^{\pm}\nu_{\mu}$, and therefore the iron yoke serves to increase the punch-through of low energy muons.

However, the muon that does penetrate the iron volume had its momentum very well measured in the tracking system, and it is only required that the track in the segmented iron volume be properly matched to the precision track from the tracking system. This matching depends on the lateral geometrical agreement of the two tracks limited by multiple scattering in the calorimeter and coil, and upon the momentum agreement. The measured momentum of the muon in the iron yoke, however, is fundamentally limited to 10%, or worse, independent of the precision of the spatial tracking measurements, independent of the momentum, and insensitive even to the depth of iron over which the track measurements are made, as shown in the following calculation, Figure 4.2.

The variation of the $\mu^{\pm}$ momentum with the sagitta, from Section 3.2.1, is

$$\delta p = p^2 \left[\frac{8}{0.3 B \ell^2}\right] \delta s \,, \tag{4.1}$$

and in this multiple-scattering-dominated iron volume, the sagitta variation has two contributions, the usual one from the chamber spatial precision ($\delta s_{\text{chamber}}$)

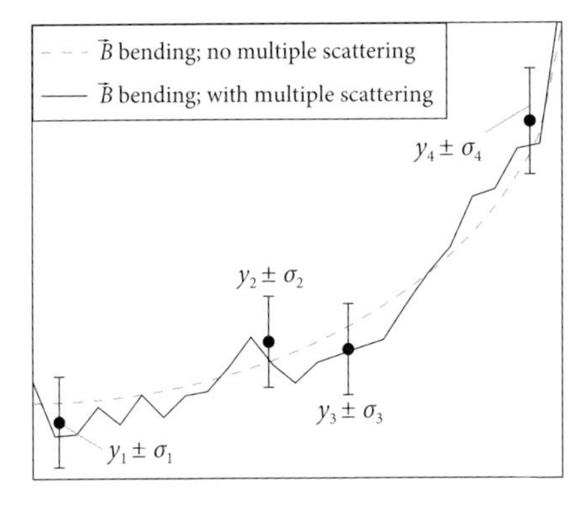

Figure 4.2 The multiple scattering and bending of a particle in a dense medium compete for curvature, and so for a $\mu^{\pm}$ in Fe the multiple scattering limits the momentum resolution to 10% independent of momentum and almost independent of the depth of the Fe.

and the second from multiple scattering given by s_{rms} in Section 3.1.2, so $\delta s = \delta s_{\text{chamber}} + \delta s_{\text{MS}}$, and where

$$\delta s_{\text{ms}} = s_{\text{rms}} = \frac{0.0136\,(\text{GeV/c})}{4\sqrt{3}\beta p}\ell\sqrt{\frac{\ell}{X_0}}\,. \tag{4.2}$$

X_0 is the radiation length of the medium (in this case, 17.6 cm for Fe) and ℓ is the depth of the Fe absorber. The $\mu^{\pm}$ is very energetic so $\beta \approx 1$, and a quick estimate of the magnitudes of these two contributions to δs, for $\mu^{\pm}$ at any energy, yields $\delta s_{\text{chamber}} \ll \delta s_{\text{MS}}$, so that $\sigma_p = \sqrt{\langle \delta_p^2 \rangle}$ and

$$\sigma_p = p^2\left[\frac{8}{0.3\,B\ell^2}\right]\cdot\left[\frac{0.0136\,(\text{GeV/c})}{4\sqrt{3}p}\ell\sqrt{\frac{\ell}{X_0}}\right]\,. \tag{4.3}$$

The radiation length of Fe is $X_0^{\text{Fe}} \approx 0.176$ m, and the maximum magnetic field is at saturation in iron, $B \approx 1.8$T, so that

$$\boxed{\frac{\sigma_p}{p} \gtrsim \frac{0.15}{\sqrt{\ell\,(\text{m})}} \quad \text{(in an iron spectrometer)}}\,, \tag{4.4}$$

a simple result independent of momentum and only weakly dependent on the depth of Fe. For a typical Fe absorber of $\ell \approx 3$ m, the $\mu^{\pm}$ momentum resolution can be no better than about 10%.

Muon identification and measurement in an iron-free detector Only one major detector is iron free, the ATLAS experiment at the LHC at CERN, as shown in Section 6.3. A high energy $\mu^{\pm}$ will penetrate the calorimeters and the solenoidal coil before entering a magnetic field region established by toroid superconducting coils, where the momentum of the muon is found by a sagitta measurement to an anticipated precision of $\sigma_p/p \approx 2\%$ for muon p_T below 200 GeV/c and a resolution of $\sigma_p/p \approx 10\%$ at $p_T = 1$ TeV/c. It will be of great interest to compare the muon physics from the CMS and ATLAS experiments in the coming years.

Muon ID by momentum balance In a dual-solenoid iron-free detector, the momentum of a $\mu^{\pm}$ from the IP will be measured in the tracking system ($p_{\mu 1}$), and any energy loss (by ionization or radiation) will be measured in the calorimeter (ΔE_{μ}) and its momentum measured again ($p_{\mu 2}$) in the annulus between the solenoids. A muon candidate must satisfy momentum balance, $p_{\mu 1} \approx \Delta E_{\mu} + p_{\mu 2}$, where the precision of this balance depends on the momentum of the muon, and the radiation of the muon inside the solenoid and its cryostat, and is typically 3 to 5%. This balance can reject pions against muons to about a factor of 30 for decay muons from W and Z^0.

Muon ID by dual readout of Čerenkov and scintillation light The separation of $\mu^{\pm}$ radiation and ionization energy losses in a dense medium has a consequence for

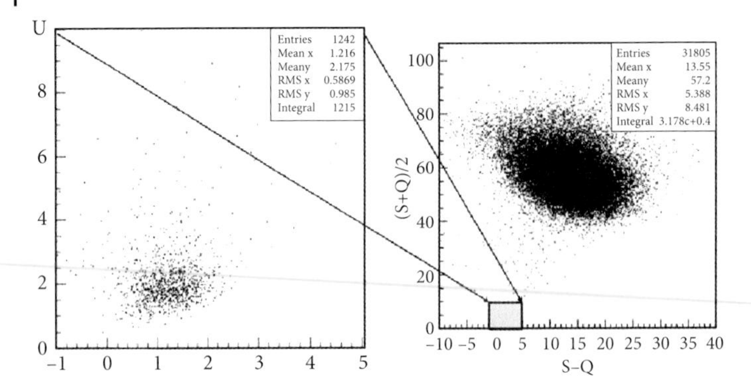

Figure 4.3 Discrimination of an 80-GeV μ^- from an 80-GeV π^- in the dual-readout DREAM module. Plot of $(S - C)$ vs. $(S + C)/2$.

$\mu^{\pm}$ identification and discrimination against $\pi^{\pm}$. In Figure 4.3 the difference between the scintillation (S) and Čerenkov (C) signals are plotted against their average for 80-GeV particles. The $\mu^{\pm}$s are well separated from the $\pi^{\pm}$s, and this remains so from the lowest measured energies (20 GeV) to the highest (200 GeV), Figures 3.8a–d.

Since the important $\mu^{\pm}$s are usually isolated, this method can be effective, whereas in a dense jet environment the calorimeter identification of the S and C signals will be difficult.

These two methods, muon momentum balance and $(S-C)$, are independent and their pion rejection factors multiplied.

4.1.2
Telling an $e^{\pm}$ from a $\pi^{\pm}$

Telling an $e^{\pm}$ from a $\pi^{\pm}$ by Depth Development

Electromagnetic showers develop in depths over 10–20 X_0 (radiation lengths) while hadrons develop in depths over 5–10 λ_{int} (nuclear interaction lengths). Referring to Figure 3.20, one X_0 is 10 to 20 times smaller than one λ_{int}, and therefore any calorimeter that measures showers in depth can separate $e^{\pm}$ from $\pi^{\pm}$ to a level limited by the probability that a π^- will charge exchange in the front of the calorimeter, producing a nearly full-energy $\pi^0 \to \gamma\gamma$ that will mimic an $e^{\pm}$. This probability is about 10^{-3}. Depth development discrimination is the main reason calorimeters are divided into "electromagnetic (em)" and "hadronic (had)" sections in depth, although this is a misnomer since about one-half of all hadrons interact in the em section.

Telling an $e^{\pm}$ from a $\pi^{\pm}$ by a Preshower Detector

In contrast to the gross division between em and had sections above, a preshower detector uses only the very beginning of a shower, perhaps the first 2 X_0 and (for Pb) 0.06 λ_{int}, and the preshower detector consists simply of this piece of Pb and a scintillator behind it. An electron will immediately begin to produce bremsstrahlung in the Pb generating a large number of low-energy γs and losing on average about 90% of its energy in 2 X_0. In particular, the γs in the MeV region have a minimum in their nuclear cross-section just below the pair-production threshold at $E_\gamma \approx 2m_e c^2$, and in this energy window they will pass easily through the Pb and Compton scatter in the scintillator, giving a large signal. On the other hand, the pion will only interact 6% of the time (in Pb) and, when it does, yield only a few *mips* of signal in the scintillator. The discrimination against $\pi^\pm$ can be large, about 100:1.

Depth development differences depend on λ_{int} being much larger than X_0 for the absorber, and higher Z is better. With reference to Figure 3.20, the ratio λ_{int}/X_0 is a monotonically increasing function of Z,

$$\frac{\lambda_{\text{int}}}{X_0} \approx 0.35 \times Z \,.$$

Therefore, U and Pb are best for $e - \pi$ separation with $\lambda_{\text{int}}/X_0 \approx 30$.

Telling an $e^\pm$ from a $\pi^\pm$ by Lateral Development

The transverse development of electromagnetic showers, like depth development, is both narrower and similar shower to shower, and these properties can be exploited in any calorimeter that is laterally segmented on a scale of 2–4 X_0. Often the Moliere radius, $R_{\text{M}} = X_0 E_{\text{s}}/E_{\text{c}}$,[32] is used to characterize the lateral spread, since typically 90% of the shower energy is contained in a cylinder of radius R_{M}. However, the core of a shower is much narrower than R_{M}, and I do not find R_{M} to be useful in calorimeter design. In any case, $\pi^\pm$-induced hadronic showers are much wider due to the $p_T \sim 200$ MeV/c transverse momentum in hadronic interactions, and the shower-to-shower fluctuations are much larger than in electromagnetic showers. Any shower width measurement with a granularity of $\sim 2X_0$ will provide $e^\pm$-$\pi^\pm$ discrimination.

Telling an $e^\pm$ from a $\pi^\pm$ by E/p

The $e^\pm$ and $\pi^\pm$ momenta will be measured in the tracking system and their energies measured in the electromagnetic calorimeter (first 20 X_0, or 1 λ_{int}). The $e^\pm$ will deposit nearly all of their energy in the calorimeter, but the $\pi^\pm$ will deposit anywhere from one *mip* (no interaction) up to $\sim$ 50% (one hadronic interaction). Therefore, $e^\pm$ will have $E/p \approx 1$ while $\pi^\pm$ will have $E/p \ll 1$, in general.

A specific and recent demonstration in precollision ATLAS data [76] of all three of the above methods, E/p and subsequent lateral and depth development selection, for e^- identification is shown by the ATLAS measurement in Figure 4.4. The e^- were highly energetic δ-rays from passing cosmic $\mu^\pm$ in the ATLAS tracking system at a depth of about 100 m at the CERN LHC.

32) $E_{\text{s}} = \sqrt{4\pi/\alpha}\, m_e c^2 \approx 21$ MeV is the so-called scale energy.

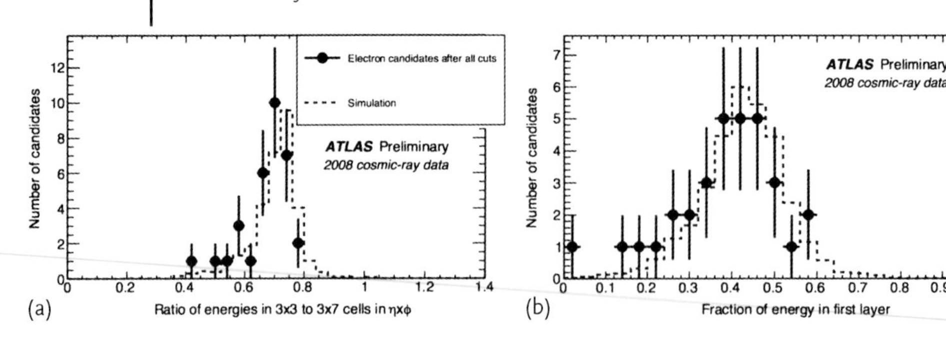

Figure 4.4 The first electrons to appear in the ATLAS detector, from the bremsstrahlung of cosmic ray muons, have been used to separate $e^{\pm}$ from $\pi^{\pm}$ and $\mu^{\pm}$ tracks using all three of the methods listed here: (a) by lateral development and (b) by depth development, preceded by an E/p selection. The dotted curves are the distributions expected from the simulation of the detector.

The cosmic $\mu^{\pm}$ sample was 3.5 M events,[33] from which about 32 e^{-} δ-rays were selected. This is a dramatic illustration of the ultra-high-energy tail to the dE/dx distribution.

Telling an $e^{\pm}$ from a $\pi^{\pm}$ by Dual Readout, vs. C

In a dual-readout calorimeter, the separate measurements of the electromagnetic part of the shower (C) and the whole shower including nuclear n and p (S) give an automatic separation of $e^{\pm} - \mu^{\pm} - \pi^{\pm}$, as seen in Figure 4.5. The calorimeters are calibrated on e^{-} for both S and C, and so $S \approx C$ for electromagnetic energy. In this fiber dual-readout calorimeter, a $\mu^{\pm}$ will deposit a fixed energy by dE/dx, plus equal signals in S and C for any radiation that takes place in the calorimeter volume. A $\pi^{\pm}$ will shower and yield the large expected fluctuations in π^{0} vs. $\pi^{\pm}$ production in hadronic interactions that are reflected in a dual-readout calorimeter as large variations in C (mostly from π^{0}s) and the partially correlated S signal as seen in Figure 4.5. The three populations of particles $e^{\pm}, \mu^{\pm}$, and π^{pm} are widely separated from each other in this plot.

Telling an $e^{\pm}$ from a $\pi^{\pm}$ by Fluctuations in $(S_k - C_k)$

The S vs. C plot of Figure 4.5 is for the whole shower, where S and C are the sums of the energies in all the channels that are believed to be activated by the particle,

$$S = \sum_{k=1}^{N} S_k \quad \text{and} \quad C = \sum_{k=1}^{N} C_k \,.$$

33) At a depth of 100 m, the mean cosmic muon energy is approximately the energy loss in 100 m of Swiss rock ($\rho \sim 3.5\,\text{g/cm}^3$), or $\langle E_{\mu} \rangle \sim 500\,\text{GeV}$, and distribution in energy continues as $\sim 1/E^2$. The CMS of $\mu^{\pm}$ radiative energy losses in a similar-sized sample range from 5 GeV to 1 TeV, Figure 3.5.

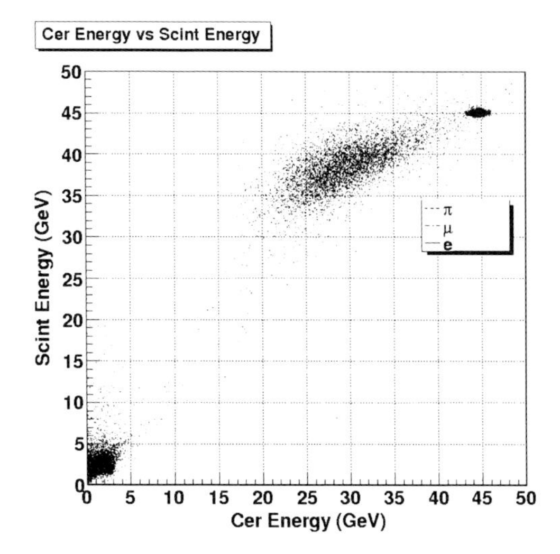

Figure 4.5 Simulated response of $e^{\pm}$, $\mu^{\pm}$, and $\pi^{\pm}$ in a dual-readout crystal and fiber calorimeter system (Chapter 7). From Di Benedetto, *et al.*, [81].

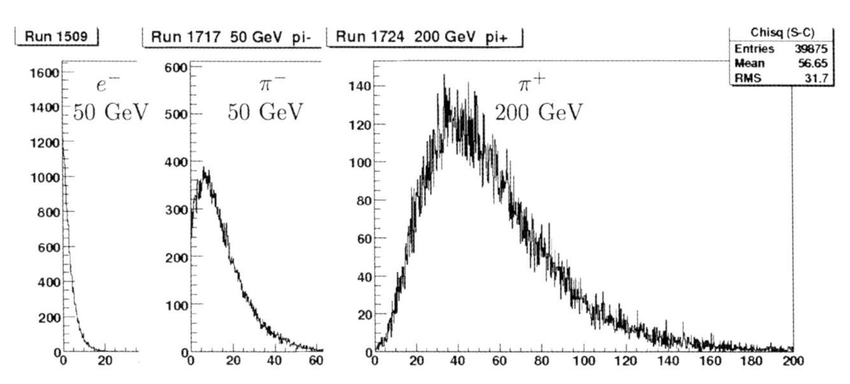

Figure 4.6 Distribution of the χ^2 for the equality of S_k and C_k signals in all the channels that comprise a shower, shown for e^- at 50 GeV, π^- at 50 GeV, and π^+ at 200 GeV.

If the shower is electromagnetic, then $S_k \approx C_k$ for all channels k. A chi-squared statistic is constructed that tests this:

$$\chi^2 = \sum_{k=1}^{N} \left(\frac{S_k - C_k}{\sigma_k} \right)^2 ,$$

where σ_k is the expected rms variation of $(S_k - C_k)$ for electrons, $\sigma_k^2 \approx 0.1\,(S_k + C_k)$. The distribution of this χ^2 is shown in Figure 4.6 for e^- at 50 GeV, π^- at 50 GeV, and π^+ at 200 GeV, with mean chi-squared values of $\langle \chi^2 \rangle \approx 1.5, 15$, and 60, respectively, for e^-, for π^- of the same energy, and for higher energy π^+. This chi-squared statistic can be widely useful in discriminating between hadronic and non-hadronic energy deposits in any dual-readout calorimeter. In the $\tau^{\pm}$ identification section, Section 4.5.3, one can discern with the naked eye that the γs are electromagnetic energy deposits and that π^- is hadronic by the channel-to-channel fluctuations in $(S-C)$. For jets and hadrons that may fluctuate to be near the $S \approx C$ point in a dual-readout calorimeter, channel-by-channel fluctuations can be used to resolve the ambiguity of hadronic vs. nonhadronic.

Telling an $e^\pm$ from a $\pi^\pm$ by the Time Development of the Scintillation Signal

Electromagnetic showers are simple and they develop in the same way, as a relativistic pancake of e^+, e^-, γ that travel transverse to the beam and at velocity $v \approx c$. Thus, the space and time structure of all electromagnetic showers is very similar. This is not the case for hadronic showers (induced by $\pi^\pm$, for example). Hadronic showers have large fluctuations in the number of $\pi^0 \rightarrow \gamma\gamma$ and the number of $\pi^\pm \rightarrow$ hadrons that are produced, and in addition there are slow neutrons ($v \sim 0.05\,c$) and spallation protons ($KE \sim 100$ MeV) from nuclear breakup. The very nonrelativistic n and p deliver ionization and scintillation light over a time period of tens of nanoseconds. The time structure of the scintillation (S) signal was measured in the SPACAL calorimeter and is shown in Figure 4.7, where the horizontal axis is the full width at one-fifth maximum of the photomultiplier signals from the scintillating fibers. The electromagnetic showers are uniform in time to 1 ns, whereas the hadronic showers fluctuate in time over 10–20 ns. In general, one cannot tell an $e^\pm$ from a $\pi^\pm$ in a tracking system (except at low momenta by dE/dx or time of flight), a calorimeter is the only way, and all techniques are based on the physical differences between electromagnetic and hadronic interactions.

Most of these $e^\pm - \pi^\pm$ discriminators are independent, and therefore their rejection factors multiple.

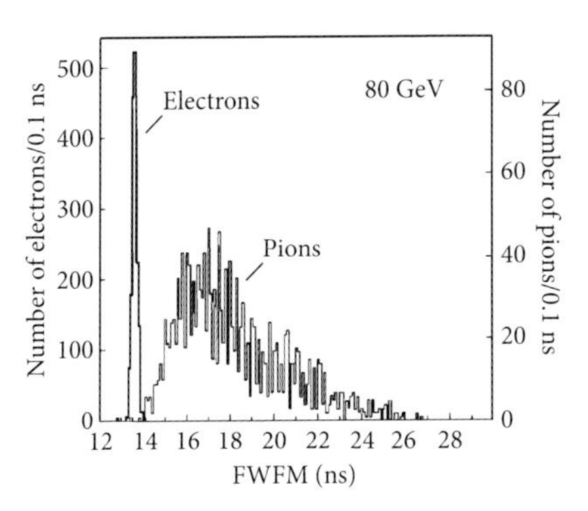

Figure 4.7 The duration of the scintillation pulse (full width at one-fifth maximum) for e^- and π^- in the SPACAL calorimeter [51].

4.2
Discriminating Hadrons from Each Other

4.2.1
Telling a $K^{\pm}$ or a p from a $\pi^{\pm}$

Telling a $K^{\pm}$ or a p from a $\pi^{\pm}$ by dE/dx

Charged hadrons are effectively separated using dE/dx measurements at momenta below 10–20 GeV/c, and the efficacy of this is shown clearly in two tracking detectors: a TPC, Figure 3.1a, and a silicon tracking chamber, Figure 3.1b, showing all the features of the Bethe–Bloch formula Eq. 3.2, in particular, the $1/\beta^2$ term at low momenta and the $\ln\gamma$ term ("logarithmic rise") at high momenta. It is expected that silicon tracking will improve as noise levels are reduced, but the "logarithmic rise" will still be suppressed by the polarization of the medium. Further improvements in gaseous tracking chambers [42] may improve on the TPC dE/dx measurement by a factor of two.

Telling a $K^{\pm}$ or a p from a $\pi^{\pm}$ by Time of Flight

The velocity (β) of a particle with known momentum depends on its mass since $\beta = p/E = p/\sqrt{p^2+m^2}$. At a collider, a prompt particle from the interaction point has a very precisely known start time (the beam crossing time, t_0) and a measurement of its arrival time after a flight path of length L is $t = L/v$, or

$$ct = \frac{L}{\beta} = \frac{L}{p}\sqrt{p^2+m^2} \approx \left[\frac{L}{2p^2}\right] m^2 ,$$

for $p \gg m$. So the mass is directly measured by the time-of-flight. For flight path $L \approx 2$ m, and time difference relative to L/c for a 1 GeV/c $\pi^{\pm}$ is 0.066 ns, and for a $K^{\pm}$ is 0.8 ns. For anticipated time resolutions near $\sigma_t \approx 10$ ps, this is interesting, but grows quadratically more difficult as p increases. It is not clear that time of flight can be used in a large-scale experiment, except for special cases, for example, at the outer edges of the detector beyond the muon system.

On the other hand, time of flight can be critically useful if massive objects are produced at the interaction point and move out into the tracking volume with very low velocity, $\beta \approx 0$, before decaying to $\beta \approx 1$ decay products. These objects could be reconstructed just like the completely invisible decay $K^0_L \to \pi^0\pi^0 \to 4\gamma$s in the KLOE experiment, Section 4.9 and Figure 4.11.

Telling a $K^{\pm}$ or a p from a $\pi^{\pm}$ by Čerenkov Light

Both the Čerenkov angle and the Čerenkov light intensity depend on particle velocity, and therefore on the mass. An excellent treatment is given in the PDG by Radcliffe [49, Section 28.5] and the most sophisticated particle identification instrument is a DIRC (Detector of Internally Reflected Čerenkov light) first used in the BABAR detector[77] at the asymmetric B factory at PEPII. Such sophisticated detectors will someday find their way into the big detectors of future colliders, but at the moment and to my knowledge, no one is thinking in these terms now. Part of the reason is that a big collider detector is often filled with tracks from jets and from beam-crossing backgrounds, and this high track density can easily defeat a DIRC, which, of course, is well designed for clean final states in $e^+e^- \rightarrow B\bar{B}$ events. A second reason might be that the physics with low energy particles is often dismissed as unimportant at a TeV collider, but as we have pointed out elsewhere, this is not a good physics argument. There will be many events with decay chains like $t \rightarrow b \rightarrow c \rightarrow s$ or $H \rightarrow b \rightarrow c \rightarrow s$ with nonzero impact parameters and lots of $K^{\pm}$ at low momenta.

For discriminating between high energy mesons and baryons, there is a small subtle difference that has been measured [50] in highly noncompensating Čerenkov fiber calorimeters. The $\pi^{\pm}$ response is larger than the p response by 20%, is asymmetric, and has worse resolution. This could be tested in a dual-readout fiber calorimeter by looking at the noncompensating ($e/h \sim 5$) signals from the Čerenkov fibers.

4.2.2 Neutral Hadrons: n and K_L^0

Neutral hadrons will interact deep in the calorimeter, and there is little hope in discriminating between them. The physics gain is small: any physics that can be done with n and K_L^0 can also be more easily done with p and K_S^0 or $K^{\pm}$. There is one weak discriminator: an n (a three-quark baryon) has a nuclear cross-section about twice as large as a K_L^0 (a two-quark meson), but the n will elastically and diffractively scatter deeper into a calorimeter. These are not big differences, but they are analogous to the measured differences between p and π^+ in a Čerenkov calorimeter, as reported by Akchurin [50] and discussed by Wigmans [29, Figure 8.27].

4.3 Identifying "Jets"

4.3.1 Discriminating a Light Quark (u, d, s) from a Gluon (g)

This is a classic problem, worked on by many people, and although a statistical separation can be made between ensembles of gluon and quark jets, no jet-by-jet algorithm or technique has been developed. The main idea for an algorithm is that gluons couple to gluons, and so a gluon-initiated jet should have a larger number of lower energy particles in it than a quark-initiated jet. A rather recent theoretical paper based on this notion and using LEP data on three-jet events, that is, $e^+e^- \rightarrow q\bar{q}g$ events in which one quark radiates a gluon, is by Eden, Gustafson, and Khoze [78]. One can usually make a good guess from the kinematic configuration of the jets which one is the gluon.

4.4
Identifying $W^{\pm} \rightarrow q\bar{q}$ and $Z^0 \rightarrow q\bar{q}$ Decays into Hadrons

The high precision reconstruction of jets and the reconstruction of $W \rightarrow jj$ and $Z \rightarrow jj$ decays with resolution on the W and Z masses of 3–4 GeV is considered crucial for any future experiment. This seems to be in reach with dual-readout calorimeters, which have a mass resolution of about

$$\frac{\sigma_{M_{jj}}}{M_{jj}} \approx \frac{34\%}{\sqrt{M_{jj}}}$$

for two-jet masses, as shown in Figure 3.24. This is the initial high precision W and Z physics in big detectors. We have gone from the barest understanding, for example, by C.N. Yang [79], of Ws and Zs to experiments in which we might measure their four-vectors to precisions similar to those of other standard model particles, such as $e^{\pm}$ and $\mu^{\pm}$. The reconstructed two-jet invariant masses for four-jet events in $e^+e^- \rightarrow W^+W^-$ and $e^+e^- \rightarrow Z^0Z^0$ events are shown in Figure 4.8 in the simulation of the dual-readout calorimeters of the 4th concept detector [75].

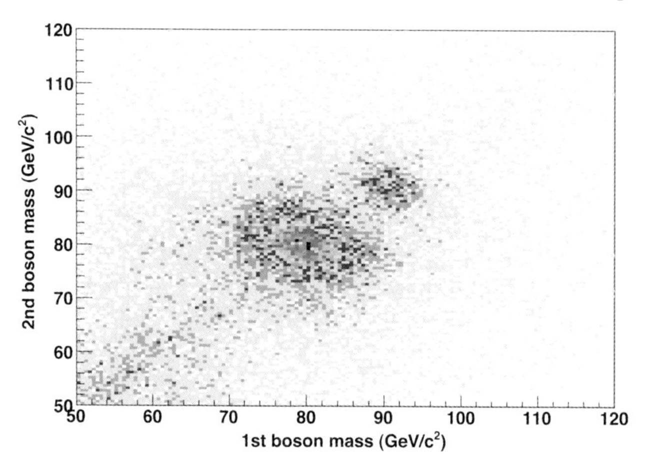

Figure 4.8 The invariant mass of two jets plotted against the invariant mass of the other two jets, in four-jet events from the processes $e^+e^- \rightarrow W^+W^-$ and $e^+e^- \rightarrow Z^0Z^0$. From Di Benedetto *et al.* [81].

4.5
Identifying Particles by Their Lifetimes

4.5.1
Identifying Weak s-baryon Decays: $\Lambda(sdu)$, $\Sigma(sqq)$, $\Xi(ssq)$, and $\Omega(sss)$

There is little interest in identifying these strange baryons in a big detector at a collider, although LHC*b* and the defunct BTeV had good reasons to study their decays. They are all easily identified and reconstructed in tracking systems with many points measured along a track, such as a TPC or a cluster-timing chamber ("continuous" tracker). These decays are not easy to find in tracking systems with a small number of high precision layers.

On the other hand, reconstructing these special baryons has physics value since their identification will be very clean and they will reveal their specific quark content.

4.5.2
Identifying Weak Heavy Quark Decays: $B(b\bar{q})$ and $D(c\bar{q})$

These weakly interacting heavy quark states have interesting lifetimes in the $c\tau \sim 100-400$ μm range, Table 2.1, and a high precision impact parameter vertex chamber can easily measure the distributions of impact parameters for uds, c, and b quark jets, as shown in Figure 3.12b. With a selection on the impact parameter significance, b/σ_b, an ensemble can be enriched in b or c jets. The only other measurement (apart from discrete reconstruction of a decay chain, for example, $B^0 \rightarrow D^+\pi^-$), is the invariant mass of the particles of the jet, and one should select relative to $m_D \approx 2$ GeV/c^2 to further separate D from B.

4.5.3
Identifying a $\tau^\pm$ Lepton

The massive and valuable $\tau^\pm$ lepton will come from the decays of W and Z, and also from the decays of a putative Higgs. In its subsequent decay, there will always be one neutrino (ν_τ) to conserve the τ-lepton number or, for leptonic decays, two neutrinos to conserve both the τ-lepton and e-lepton or μ-lepton number. Therefore, any event with a $\tau^\pm$ in it will have missing momentum, and because the τ is low-mass, $m_\tau \approx 1.8$ GeV/c^2, the missing neutrino momentum vector will be nicely aligned with the charged tracks of the $\tau^\pm$. In the language of kinematic constraints, there is only one unknown, the magnitude of the total neutrino momentum, not three unknowns. There are four ways to identify a likely $\tau^\pm$ lepton:

- *Identifying a $\tau^\pm$ Lepton by lifetime (impact parameter)*: The $\tau^\pm$ lifetime is $c\tau \approx 87$ μm, and it can be tagged by impact parameter and further tagged by its isolation from other objects in the event since it decayed from a W or a Z.
- *Identifying a $\tau^\pm$ Lepton by topology*: 85% of $\tau^\pm$ decays are topologically one-prong ($e^\pm, \mu^\pm$, or $\pi^\pm$) and the other 15% are three-prong decays. Both of these are topologically distinctive, especially since the $\tau^\pm$ mass is only 1.8 GeV/c^2 and the tracks of the three-prong will all be close to each other in angle.
- *Identifying a $\tau^\pm$ Lepton by exclusive reconstruction of $\tau^\pm \rightarrow \rho^\pm\nu_\tau$*: A specific final state with physics interest is the decay to a spin-1$\hbar$ ρ meson, which decays immediately to $\pi^\pm\pi^0 \rightarrow \pi^\pm\gamma\gamma$, and so the detector final state is

$$\pi^\pm\gamma\gamma\,,$$

Figure 4.9 Simulated $\tau^- \to \rho^- \nu_\tau \to \pi^- \gamma\gamma$ event in dual-readout crystals and fiber calorimeter system.

a particularly simple final state of one charged pion and two photons, shown in Figure 4.9. The vertex is in the upper right-hand corner, the π^- track is downward to the left.

The simulated signals from the BGO crystal calorimeter are plotted in $\theta-\phi$ coordinates in Figure 4.10a for the scintillation pulses and Figure 4.10b for the Čerenkov pulses. A close examination of these pulse heights shows that $C \approx S$ for the total shower and also that $C_k \approx S_k$ *channel by channel*. These are essentially a positive identification for an electromagnetic object as discussed in Section 4.1.2. Furthermore, the two γs from this 18.5-GeV π^0 are separated spatially in the fine-grained crystal calorimeter. The individual γ energies are 3.1 and 15.4 GeV.

The π^- does not interact in the 1 λ_{int} crystal section and deposits all of its energy in the deep dual-readout fiber section. The fiber scintillation signals (S) are shown in Figure 4.10c in channels of size $\Delta\theta \approx \Delta\phi \approx 1.6°$ and the corresponding fiber Čerenkov (C) signals in Figure 4.10d. A close examination shows that $S > C$ for the total and also that $S_k - C_k$ fluctuates *channel by channel*. The charged track that points to this hadronic shower has a measured momentum of 73 GeV/c, penetrates the BGO crystals without interacting, and is clearly identified as hadronic, most likely a π^-. This level of detail for a single event is a measure of the strength of a finely laterally segmented dual-readout calorimeter.

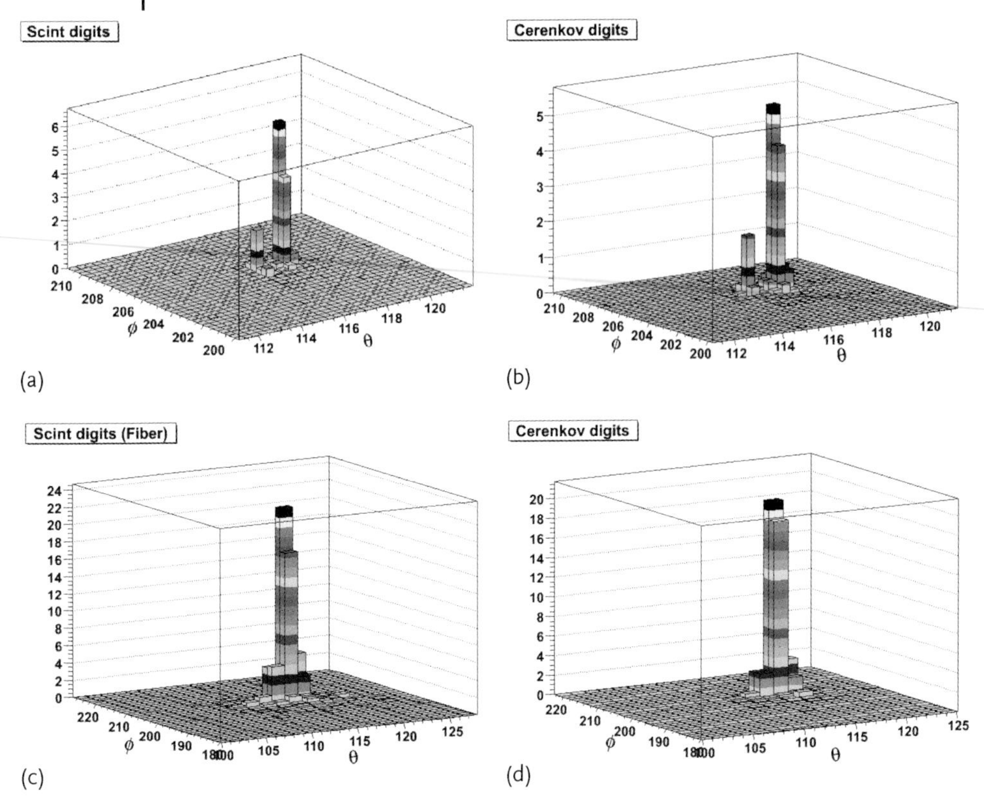

Figure 4.10 Scintillation and Čerenkov signals from a $\tau^- \to \rho^- \nu_\tau \to \pi^- \pi^0 \to \pi^- \gamma\gamma$ decay. (a) The scintillation signals in the BGO crystals, and (b) the Čerenkov signals in the BGO crystals. The observation that $S_k \approx C_k$ *channel by channel* confirms these narrow showers as electromagnetic. The BGO channels subtend a solid angle given by $\Delta\theta \approx \Delta\phi \approx 0.4°$. (c) The scintillation signals in the deep fiber calorimeter, and (d) the Čerenkov signals) in the fiber calorimeter. The observation that $S_k - C_k$ fluctuates *channel by channel* is an indication that this shower is not electromagnetic. The fiber channels subtend a solid angle given by $\Delta\theta \approx \Delta\phi \approx 1.6°$.

4.6 Telling a γ from $\pi^0 \to \gamma\gamma$ and $e^\pm \to e\gamma$

A single photon can be tagged as electromagnetic and laterally narrow if the channel segmentation is fine enough, typically 1–2 X_0 transverse. The primary physical background to a single photon is the decay $\pi^0 \to \gamma\gamma$. If the lateral granularity of the detector is $\Delta\theta$ and the π^0 is at high energy so that $m_{\pi^0}/E_{\pi^0} \leq \Delta\theta$, then these two photons can mimic a single photon. A less likely way to mimic a single photon is for the π^0 to decay very asymmetrically, $\theta^* \approx 0$, so that the p_T of both photons is near zero.

Depth development in a preshower detector is another discriminant since a single photon mean free path is $9/7\ X_0$ and less likely to start a shower, whereas two photons have twice the probability of starting showers. Neither of these is a clear discriminator.

4.7 Identifying a Neutrino (ν) in an Event

A single neutrino, ν, for example, from the decay of a $W^{\pm} \rightarrow e^{\pm}\nu_e$, will not interact in a detector, and the only hope is to assume that just one ν is missing from the event and to use all other momentum measurements ($\boldsymbol{p}_i$) to estimate the ν three-vector as

$$\boldsymbol{p}_\nu = -\sum_i \boldsymbol{p}_i .$$

The precision on this sum is the sum in quadrature of all the terms and would in general be limited by the worse measurement in the event, for example, one high momentum track or one energetic jet. This is a further reason to achieve excellent hadronic energy resolution.

4.8 Transition Radiation Proportional to γ

Transition radiation is generated by a charged particle passing between two media with different dielectric constants, so that the electric field suffers a discontinuity at the boundary that is resolved by emitting a photon. In practice, many boundaries are needed to get a sufficient number of detectable photons, which are in the X-ray region, and the radiated energy is proportional to γ of the particle, which is very interesting for particle identification. In spite of this feature, this technique is difficult to implement because the photons are produced very forward and lie on top of the charged particle that produced them.

4.9 Time to Mass

There is one beautiful example of K_L^0 identification by time of flight on the decay photons from the decay π^0s of the K_L^0. Since I suspect this may have a useful parallel for very massive objects at big colliders, two figures from a KLOE [80] paper are shown in Figure 4.11. Figure 4.11a is a simple schematic of the geometry for a photon that decays from a long-lived K_L^0 in the chamber volume and hits the calorimeter where its time of flight is measured to a spectacular precision of $\sigma_t \approx 54\,\text{ps}/\sqrt{E} \oplus 140\,\text{ps}$. The K_L^0 flight path is measured by a single photon since the initial state kinematics is known; more than one photon overconstrains the K_L^0 flight path and defines its trajectory in three dimensions.

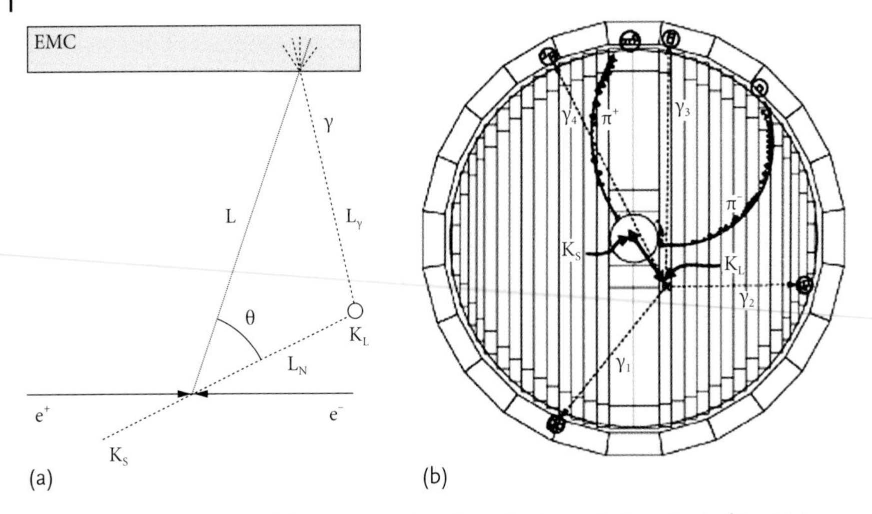

Figure 4.11 (a) Schematic of the geometry of a γ from the decay of a long-lived K^0K_L. (b) A KLOE event with a $K_S^0 \to \pi^+\pi^-$ very near the interaction point, and $K^0K_L \to \pi^0\pi^0 \to \gamma\gamma\gamma\gamma$ displaced from the IP.

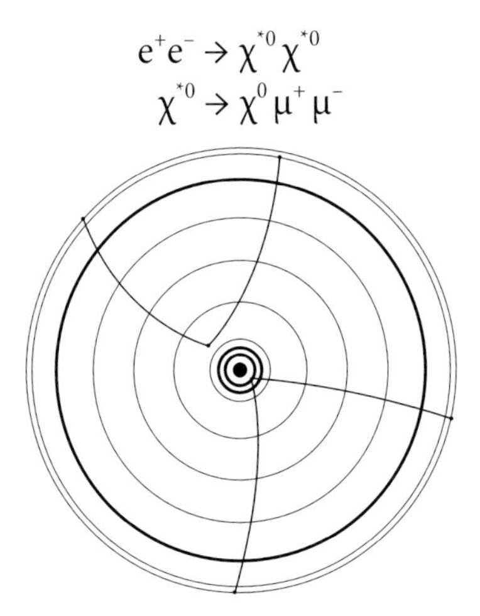

Figure 4.12 The production and decay of a massive SUSY particle and measurement of its mass by time of flight.

The KLOE event in Figure 4.11b is $e^+e^- \to K^0\bar{K}^0 \to K_S^0K_L^0$, and at this energy the K_L^0 has a velocity of $\beta \approx 0.21$. Therefore, for a flight path of only a few centimeters, the time delay before the emission of the prompt photons from the two π^0s is very measurable.

Many speculative massive objects have been suggested over the years that would decay into nearly massless particles such as all three standard model leptons, $\tau^\pm, \mu^\pm$, and $e^\pm$, or that might decay through a $W^\pm$ or Z^0 that in turn decays into nearly massless particles. In any case, for a putative particle in the 1 TeV/c^2 range and with a commensurate long lifetime, its velocity, and therefore its mass, can be measured from the times of flight of its decay products. An example is shown in Figure 4.12 of the production of a massive supersymmetric (SUSY) particle that decays to a least-SUSY particle (LSP), which is noninteracting and missing from the detector, plus near-massless particles such as $\mu^+\mu^-$:

$$e^+e^- \to \chi^{0*}\bar{\chi}^{0*} \to \chi^0\mu^+\mu^-\bar{\chi}^0\mu^+\mu^- .$$

In this case, the time of flight on the four muons will uniquely measure the mass of the χ^{*0}.

4.10 Problems

1. In the few-GeV region, both specific ionization, dE/dx, and time of flight discriminate between particles of different mass. Assume a dE/dx resolution of 3.0% in the gaseous medium of Figure 3.1a. Assume a time-of-flight resolution of 10 ps over a flight path of 1 m. What is the difference, in standard deviations, between a $\pi^\pm$ and a $K^\pm$ at 5 GeV/c in each system?

2. Is it useful, from a physics point of view, to be able to discriminate between light quarks (uds) and gluons (g) in an experiment at an e^+e^- collider? At a proton collider?

3. Apply the KLOE idea (Section 4.9) to a putative massive particle that decays into muons, $M \to \mu^+\mu^-$, and is pair-produced at the Fermilab Tevatron, $\sqrt{s} \approx$ 2 TeV. The event has four muons, $\mu^+\mu^-\mu^+\mu^-$. The event sample consists of six such events, and in each event two muons come from one decay, and two muons from another decay. The muons arrive at the calorimeters within a few nanoseconds of each other and with approximate times of 140 ns, 50 ns, 70 ns, 230 ns, 20 ns, and 100 ns. The sum of the momenta of the four muons in each event is about 800 GeV/c^2. What is the mass and mean lifetime of this putative massive particle?

4. The maximum energy that can be given to an atomic electron by a high energy particle depends on its mass and momentum, m, p, given in the PDG [49, Section 27.2] as

$$T_{\max} = \frac{2m_e c^2\beta^2\gamma^2}{1 + 2\gamma m_e/m + (m_e/m)^2} . \tag{4.5}$$

   Feel free to approximate this expression and suggest a way to use this for $\pi - K - p$ discrimination if these δ-rays could be measured.

5. To make Problem 4 practical, one needs to be able to measure the energies of δ-rays. In the bubble chamber photographs in Chapter 1, this is easy by

inspection. Suggest ways to do this in a gaseous detector (such as a TPC) and in a silicon detector.

6. Are there exceptions to the statement that particle identification requires two, or more, measurements?

7. It is possible to measure the momentum of a low energy particle in an emulsion by measuring the scattering angles of the track and extracting θ_{rms} to get p. In the far forward region of a collider detector, a muon travels through the iron and about parallel to $\boldsymbol{B}$, and there is no bending. Up to what momentum can this method be used to measure the muon momentum to 10% or better? Assume any number of infinitely precise spatial measurements throughout the iron.

8. What are the fundamental limitations on the following particle identifications due to the effects stated:

   a) telling an e^- from a π^-: due to $\pi^- N \rightarrow \pi^0 N$ in the first collision in the electromagnetic calorimeter such that the π^0 energy is within $2\sigma_E$ of what would be the calorimeter-measured e^- momentum;
   b) telling a $\mu^\pm$ from a $\pi^\pm$: due to the decay-in-flight $\pi^\pm \rightarrow \mu^\pm \nu$ in which the decay momentum of the $\mu^\pm$ is within 20% of the tracking-measured $\pi^\pm$ momentum (that is, within a $2\sigma_p$ acceptance for the momentum resolution of an iron-based muon system);
   c) telling a single γ from a $\pi^0 \rightarrow \gamma\gamma$: due to the spatial overlap of the two γs in a laterally segmented electromagnetic calorimeter of $\Delta\theta$;
   d) telling a single γ from an $e \rightarrow e\gamma$: due to a hard bremsstrahlung in the vertex chamber in which the electron is left with only a few GeV of energy and the γ takes 90% of the energy.

9. Jets are particularly challenging if you need to tag their identity, for example, from a b or c quark. The first might be a vertex impact parameter tag with probabilities shown in Figure 3.12 for the decay tracks from D and B decay. A second might be to test for semileptonic decays of D and B in which either a $e^\pm$ or $\mu^\pm$ is identified within the jet. In this semileptonic case, there must also be a missing ν so that the overall event will not exactly balance momentum. Examine all three of these cases, and estimate their relative efficacies.

10. Some day someone must find a means to distinguish between light quark uds jets and gluon g jets. A physics generator (for example, Pythia) can give you a large sample of both, with big fluctuations in the number of charged particles and their momenta from event to event, and fluctuations in jet shapes, and so on. Discuss the characteristics of a big detector that might be able to make measurements that could separate uds from g jets. The physics gain could be huge, especially at a hadron collider.

5
Particle Accelerators and Colliders

I recall the day when I had adjusted the oscillator to a new high frequency, and, with Lawrence looking over my shoulder, tuned the magnet through resonance. As the galvanometer spot swung across the scale, indicating that protons of 1-MeV energy were reaching the collector, Lawrence literally danced around the room with glee. The news quickly spread through the Berkeley laboratory, and we were busy all that day demonstrating million-volt protons to eager viewers.

– M. Stanley Livingston, January 1932

It was Rolf Wideröe who started it all with a small linear accelerator working in the radio-frequency (RF) range,[34] and Lawrence had the simple, yet brilliant, idea that if the particles were bent around in a circle, they could reuse the Wideröe linac again and again. In some sense, accelerator physics has been a repeat of these ideas, in much the same way that particle physics has been a repeat of the Rutherford scattering experiment, in which the electrostatic accelerator was a nucleus. Rutherford had the foresight, however, to commission Cockroft and Walton to build an electrostatic particle accelerator, now universally called a Cockroft–Walton, which is essentially a large electrostatic potential difference, V_0, across which the desired particle is accelerated, gaining a kinetic energy of $K = V_0$ electron volts (eV).

Accelerators are primary and essential in the pursuit of high energy physics through particle-particle scattering. There are many surprisingly excellent and readable books on accelerators, Edwards [82], Reiser [83], Wille [84], Persico [85], Conte [86], Wilson [87], and Chao and Tigner [88], the classic Bruck [89], in addition to a semipopular and interesting book by Sessler and Wilson [6] that compactly displays the big picture over one century.

34) His advisor told him to stop working on it, since it was sure to fail. Wideröe published his idea in *Archiv für Elektrotechnik*, which Lawrence could not read in the German, but the figures were clear enough.

35) Some have said this was accidental, and that only after the fact were the focusing properties, both electrostatic and magnetic, of the cyclotron Ds and its magnetic fringe field understood.

Particle Physics Experiments at High Energy Colliders. John Hauptman

ISBN: 978-3-527-40825-2

5.1 Cyclotrons, Betatrons, Synchrotrons, and FFAGs

The beam in Livingston and Lawrence's cyclotron was small (about 1 mm in diameter) and stable.[35] The imperfect magnetic fringe field of the cyclotron, shown in Figure 5.1, provided a simple nearly linear vertical restoring force for vertical excursions from the median plane, resulting in simple harmonic motion in the vertical (y) plane since the force on a beam particle is $\boldsymbol{F} = e\boldsymbol{v} \times \boldsymbol{B}$. A particle that makes an excursion up (or down) from the median plane finds itself in field with a radial component B_{r} that provides a vertical restoring force down (or up).

Horizontal (x) focusing is also due to a linear restoring force from the slow falling off of the axial bending field ("weak focusing"), B_z, as

$$B_z(r) \approx B_0 \left(\frac{r}{r_0}\right)^{-n}, \quad \text{where } n \text{ is the "field index": } 0 < n < 1\,.$$

The range of n ($0 < n < 1$) turns out to be exactly the condition for both axial and radial stability, that is, the bending field must be weaker for particles that make a radial excursion outward from the central (design) orbit. The restoring force is linear for small excursions and the oscillations simple harmonic in the (r, z) coordinates of beam particles. They are called betatron oscillations because they were understood and formulated in the context of the first electron synchrotron, or betatron, in which the accidental fringe field of the cyclotron was deliberately designed to particular n-values within $0 < n < 1$. The field index can be written as

$$n = -\frac{r}{B_z}\left(\frac{dBz}{dr}\right)_{z=0}$$

and if the field index $n < 1$, then the radial force is restoring and the beam is focused in r. If $n > 0$, then the vertical force is restoring and the beam is focused vertically in z. For field indices n outside of the range (0,1), the oscillations grow exponentially in either r or z, and the beam is unstable. In a betatron, field indices

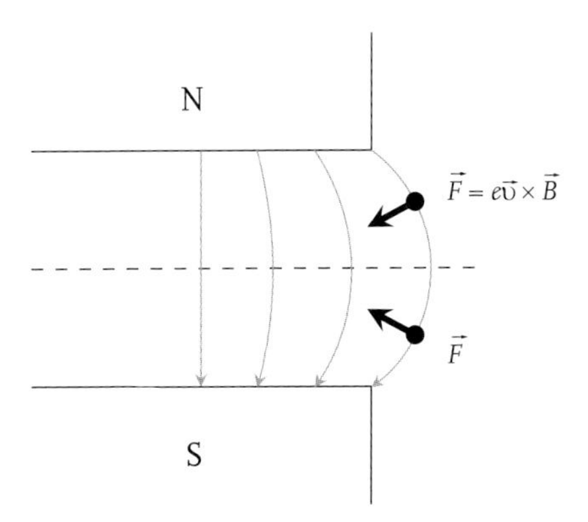

Figure 5.1 Magnetic fringe field of a cyclotron showing the vertical focusing.

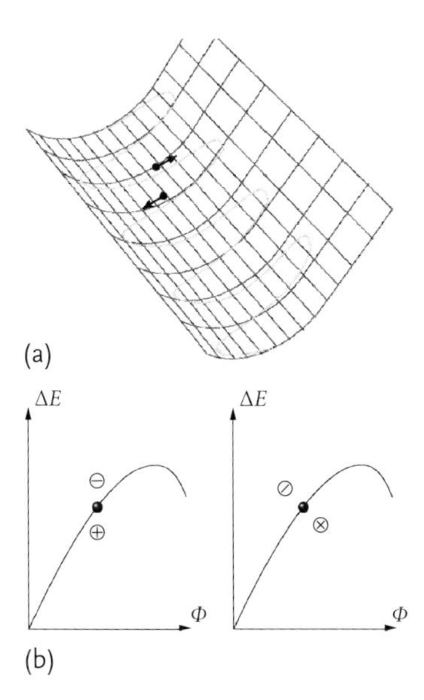

Figure 5.2 Betatron oscillations (a) and energy oscillations (b).

are usually chosen in the range $n \approx 0.65$–0.70 so that vertical focusing is stronger than horizontal focusing and the beam pipe can be a flattened ellipse with more radial space for synchrotron oscillations on top of the betatron oscillations.

Particles in a Bunch Are Stable in Three Dimensions (Figure 5.2)

Particle accelerators focus beams with quadrupole magnets among the bending dipole magnets of the machine lattice, and for a small deviation of a particle from the equilibrium orbit (the perfect orbit through the center of the lattice), the Lorentz force $\boldsymbol{F} = e\boldsymbol{v} \times \boldsymbol{B}$ provides a linear restoring force, and the differential equations for (x, y) deviations are

$$x'' + K_x(s) = 0 \quad \text{and} \quad y'' + K_y(s)y = 0 \, .$$

where s is the coordinate going around the ring on the equilibrium orbit. For K_x and K_y constants, the motion is simple harmonic oscillation about the equilibrium orbit, and approximately harmonic since $K_x(s)$ and $K_y(s)$ are not constant. The general solution for $x(s)$ resembles an oscillator

$$x(s) = A\sqrt{\beta(s)}\cos(\psi(s) + \delta) \, ,$$

where $\beta(s)$ is called the β-function of the machine and related to the amplitude of the betatron oscillation at any point in the machine. At the interaction point where very small beams are required for high luminosity, the value of β-function is denoted as β^*. Momentum excursions from the equilibrium orbit are called synchrotron oscillations, and a bunch is phased with the RF acceleration field, E, so that a particle below design momentum arrives when E is higher, and a particle above design momentum arrives at a lower E field. On the next turn around the machine, both particles are closer to the ideal orbit.

The rather weak magnetic fields of cyclotrons and betatrons are replaced by strong quadrupoles (Figure 5.3) in the focusing fields of synchrotrons and, because a quadrupole field focuses (F) in one coordinate while defocusing (D) in the other, quadrupole doublets $F_x D_x$ (and triplets $F_x D_x F_x$) are used as net focusing elements.

A particle undergoes stable oscillations in x, y guided entirely by magnetic elements through which particle energies remain exactly constant. The particles are accelerated in the longitudinal coordinate z by a time-dependent electric field usually in the RF range and the idea of phase stability in a synchrotron is that particles with too low an energy arrive at an RF point in phase with a larger electric field, and, conversely, particles with too high an energy arrive to find a lower electric field. This "restoring force" in phase or space is also almost linear, leading to almost simple harmonic energy oscillations of the particles about the momentum of the design orbit.

A cross between a cyclotron and a synchrotron is a fixed-field alternating-gradient (FFAG) machine is now receiving renewed interest after several decades. New designs are emerging and proponents argue they would be ideal for high-flux and high-rate protons for proton therapy, cargo inspection, and as the front end to a Muon Collider or neutrino factory.[36)]

5.1.1 Beam Rates, RF, Machine Geometry

The RF cavities that provide the acceleration also bunch the beams longitudinally. As seen in the sidebar figure, a particle ahead of the bunch will be accelerated less, and this restoring force is linear as long as the bunch resides in a nearly linear region of the RF wave form. In fact, the bunch should be about one-tenth of an RF wavelength, and hence there is a relationship between the bunch length and the RF. Since $c = f\lambda$, and taking the bunch length to be one-tenth of a full waveform, $\ell_{\text{bunch}} \approx 0.1\lambda$, the product of the RF frequency and the bunch length should be $0.1c$, or

$$f\,(\text{MHz})\,\ell_{\text{bunch}}\,(\text{cm}) \approx 0.1c \approx 3000\,\text{cm MHz}\,.$$

Current and planned machines have values of 3200 cm MHz (LHC), 2500 cm MHz (Tevatron), and 1600 cm MHz (HERA). Electron machines require more RF to compensate for synchrotron radiation, Section 5.3.1.1, so this constant is usually

36) Fermilab Workshop, 21–25 September 2009. http://conferences.fnal.gov/ffag09/.

smaller because the beam size in z is smaller: 500 cm MHz (LEP), 500 cm MHz (PEP), 400 cm MHz (KEKB), 600 cm MHz (CESR), 800 cm MHz (VEPP-4M), and 500 cm MHz BEPC-II).

The number of bunches that can be put into an RF train of pulses around a machine is proportional to the RF frequency, but not every RF "bucket" is filled. There is a wide range of beam rates, that is, bunch crossing intervals, that depend on the circumference of the machine, the RF frequency, and the bucket filling factor. Many machines and their operating parameters are listed in the PDG [49, Section 26].

5.1.2
Machine Backgrounds

The simplest and most fundamental difference between electron and proton machines is that electrons radiate when accelerated by the magnets of the accelerator, and protons do not. Therefore, a proton bunch that fills a spatial volume of $\delta x \delta y \delta z$ within which the protons have a momentum spread of $\delta p_x \delta p_y \delta p_z$ will have a six-dimensional "phase space" volume of $\delta x \delta y \delta z \delta p_x \delta p_y \delta p_z$ that remains constant. In contrast, electrons will quickly radiate away their $\delta p_x \delta p_y$ components, the bunch will reestablish equilibrium, and quickly the phase space volume will shrink down to a volume mainly limited by the quantum fluctuations in the radiation process. A hot proton bunch remains hot; a hot electron bunch cools.

Thus, after a proton beam is injected, it must be "scraped" to remove the protons with large undamped amplitudes.

It is quite usual that the backgrounds from a machine are not understood when the machine turns on, and it is fortunate that most machines usually turn on at a small fraction of their design intensity. At the LHC, for example, initial running is at 1% of design, allowing the detectors enough time to understand and ameliorate the already well-studied backgrounds. There are many possible sources of machine background. Particles can get outside the stable three-dimensional volume of the lattice, after which interactions with the beam elements and other materials result in a wall of particles coming down the tunnel at the detectors. The residual air in the beam pipe, however small, is a constant source of "beam-gas" interactions and particle backgrounds in the detector. These are so particular to each machine that no common rules are available, and experience is usually a good guide.

Mature machines like the Fermilab Tevatron are so "clean" that experimenters only need to turn off high voltages during p and $\bar{p}$ injection and scraping, and within minutes the voltages are raised and data taking resumed. Machines with positive beams like the CERN Intersecting Storage Rings (ISR) actually "pump the vacuum" by electrostatically driving positive ions of N_2^+ and O_2^+ into the cold beam pipe walls from which they emerge.

Electron machines are usually dominated by the radiation from the electrons that are accelerated by the magnetic lattice elements. For a circular machine (PEP, TRISTAN, LEP) electrons radiate in the direction of the detector in the final bending arcs that are several meters away from the detctor. In a linear collider for which the beams must be focused to very small transverse dimensions, the backgrounds come from the crossing of the two beams in which the electromagnetic fields of one beam on the particles of the other beam are huge, leading to what has been named "beamsstrahlung". At the ILC [1], approximately 3 TeV of electromagnetic *debris* is generated in a small forward cone during each bunch crossing, and it would be larger at CLIC [2], where the beams are smaller and the energies higher.

5.2 Beam Optics

The guiding and focusing of charged particle beams by magnetic fields begins with the Lorentz force

$$F = q\boldsymbol{v} \times \boldsymbol{B} \, ,$$

and because the force is perpendicular to the particle velocity at all times, the work done on the particle is zero, and all particles remain at fixed energy. For high momentum particles whose radii of curvature in the magnetic fields are large compared to the sizes of the magnets, a simple linear treatment of particles traversing dipoles, quadrupoles, and solenoidal magnetic fields, in addition to field-free regions between magnetic elements, are illustrated here. A charged particle in a beam or a machine lattice moves along the z-axis of the magnetic channel, and its transverse position and slope are x_1 and $x_1' \equiv dx_1/dz$. The simplest formulation of beam optics is in terms of transfer matrices M that specify the new (x_2, x_2') at the exit from the magnetic element as a function of the incoming particle, (x_1, x_1'),

$$\begin{bmatrix} x_2 \\ x_2' \end{bmatrix} = M \begin{bmatrix} x_1 \\ x_1' \end{bmatrix} .$$

This simple "thin lens" formulation is linear, ignores complications of fringe fields and imperfect magnets, and ignores path-length difference through the magnetic elements. Nevertheless, the principles are the same as an exact solution from the equations of motion. Examples of transfer matrices are as follows.

Drift Section length ℓ, $B = 0$.

The particle slope does not change ($x_2' = x_1'$), but its new position after a distance ℓ becomes $x_2 = x_1 + x_1' \cdot \ell$. Therefore, the transfer matrix M_{drift} is defined by

$$\begin{bmatrix} x_2 \\ x_2' \end{bmatrix} = \begin{bmatrix} 1 & \ell \\ 0 & 1 \end{bmatrix} \begin{bmatrix} x_1 \\ x_1' \end{bmatrix} \quad \text{drift section, length } \ell \, .$$

Quadrupole $B_x(0, y) = gy$ and $B_y(x, 0) = gx$, $g =$ field gradient.

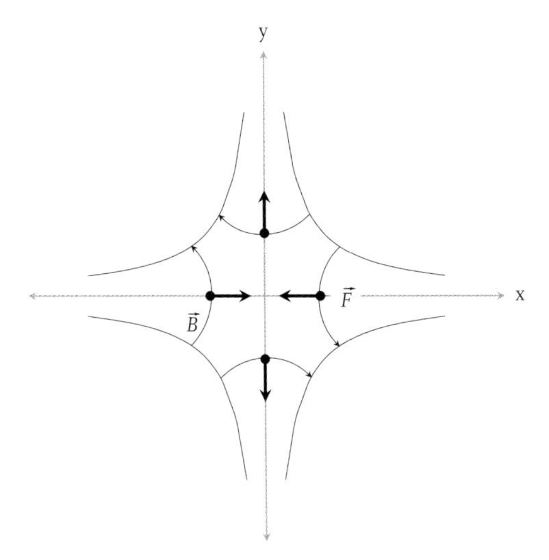

Figure 5.3 A quadrupole that is focusing horizontally and defocusing vertically.

A quadrupole field can be defined by hyperbolic (infinitely permeable) iron pole-face surfaces given by $xy = a^2/2$, where a is the aperture radius of the quadrupole, as shown in Figure 5.3. The $\boldsymbol{B}$ field lines go from the north pole to the south pole. In the thin lens approximation, a quadrupole lens does not change the position of the particle, $x_1 = x_2$, but does change the slope depending upon its distance from the axis as $x_2' = x_1' + gx_1$, where g is the spatial field gradient of the quadrupole, $g = dB/dr = B_0/a$, where B_0 is the field at the pole-face iron surface. The gradient is related to the focal length of the lens as $g = -1/f$. Therefore, the transfer matrix M_{quad} is defined by

$$\begin{bmatrix} x_2 \\ x_2' \end{bmatrix} = \begin{bmatrix} 1 & 0 \\ 1/f & 1 \end{bmatrix} \begin{bmatrix} x_1 \\ x_1' \end{bmatrix} \quad \text{quadrupole, gradient} = -1/f\ .$$

If a quad is focusing in one coordinate, it is defocusing in the other coordinate and, in order to achieve focusing in both coordinates, two quadrupoles of opposite sign are put together, a quadrupole "doublet". Better is a quadrupole "triplet" with half-strength quads of one sign on the ends and a full-strength quad of the other sign in the middle. Both of these configurations achieve net focussing in both the x and y planes.

Dipole length = ℓ, uniform field B_y, radius of curvature ρ.

$$\begin{bmatrix} x_2 \\ x_2' \end{bmatrix} = \begin{bmatrix} 1 & \rho\sin(\ell/\rho) \\ 0 & 1 \end{bmatrix} \begin{bmatrix} x_1 \\ x_1' \end{bmatrix}$$

dipole, length ℓ, radius of curvature ρ

Solenoid Axial field B_z, length ℓ.

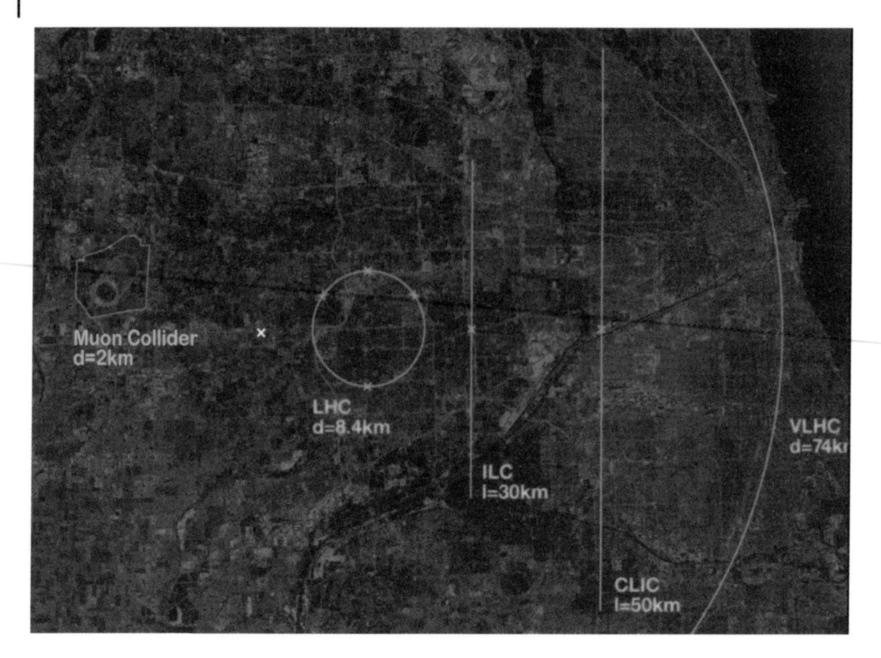

Figure 5.4 Relative geographical sizes of a Muon Collider within the Fermilab site, the LHC, ILC, CLIC, and a Very Large Hadron Collider (VLHC) superposed on the Chicago area from Lake Michigan to Fermilab.

A particle with momentum exactly along z, $\boldsymbol{p}||\boldsymbol{B}$, will experience no deflection. Transverse momentum components, p_x and p_y, result in helical orbits since x and y are not independent, and so a solenoid requires a four-by-four transfer matrix:

$$\begin{bmatrix} x_2 \\ x_2' \\ y_2 \\ y_2' \end{bmatrix} = \begin{bmatrix} 1 & \sin z\ell/\ell & 0 & (1-\cos z\ell)/\ell \\ 0 & \cos z\ell & 0 & \sin z\ell \\ 0 & (-1+\cos z\ell)/\ell & 1 & \sin z\ell/\ell \\ 0 & -\sin z\ell & 0 & \cos z\ell \end{bmatrix} \begin{bmatrix} x_1 \\ x_1' \\ y_1 \\ y_1' \end{bmatrix} \quad \text{solenoid, } B_z\ .$$

Such solenoids are used in low energy β-ray spectrometers and also, recently, in collection and cooling channels of a Muon Collider.

In the simplest sense, the magnetic guide field of an accelerator is just a collection of the necessary beam optics elements, and this is approximately true for linear accelerators. For a circular machine, a critical condition is that the optics "match" after one turn, or after an integral fraction of one turn.

5.3
Detectors at Electron, Proton, and Muon Colliders

"...smashing two Swiss watches together to figure out how they work."
– Richard Feynman, describing proton–proton machines

Feynman's comment is famous, but no one yet has remarked that sometimes electron machines are like stirring up a hornet's nest to find the queen. It is most often electromagnetic *debris* that trips the high voltages on wire chambers and drives up the occupancy of all tracking chambers.

The characteristics of detectors for electron (e^+e^-) and proton (pp or $\bar{p}p$) machines differ in almost every respect, directly leading to very different detectors at these machines. Electron machines are generally considered easier than proton machines at equivalent "physics" energies, that is, similar ee and qq center-of-mass energies, chiefly because there are fewer particles in each collision, fewer collisions per second, and no "beam remnants" or "underlying event" coming from the soft nonperturbative interactions of the spectator quarks in protons that are not part of the energetic collision.

Ideally, a detector should complete its measurements of all the particles of an event before the next bunch crossing or event arrives.

A comparison of the sizes of a Muon Collider (on the fermilab site), the LHG, ILC, CLIC, and a very large hadron collider (VLHC) is shown in Figure 5.4.

5.3.1
Electron Colliders: e^+e^-

All electron colliders to date have been circular, the most recent being the TRISTAN collider at KEK that ran at 60 GeV, an unfortunate energy limited by the geographical size of the KEK accelerator site.[37] The most successful electron machine was the Large Electron Positron (LEP) collider with e^+e^- collisions from the Z^0 at $\sqrt{s} \approx 90$ GeV to above the W^+W^- threshold at 200 GeV. The LEP collider had four equally excellent detectors, ALEPH (Apparatus for LEP Physics), DELPHI (Detector for ELectron, Photon and Hadron Interactions), OPAL (Omni Purpose Apparatus for LEP), and L3 (situated in LEP interaction region number 3). The most daring collider was SLC (SLAC Linear Collider) that ran at Z^0 and was, simultaneously, the prototype linear collider fed by the SLAC linac at 120 Hz, or 8.3 ms bunch crossing time.

The two planned electron machines, ILC and CLIC, are both linear colliders and have bunch crossing times of 337 and 0.4 ns, respectively. This is a huge difference for a detector. The ILC interval is well matched to most fast detectors, such as silicon strips and pixels, and all optical detectors, in addition to being a comfortable interval in which to collect the neutrons from nuclear breakup in the calorimeters.

37) In Japan, land is owned down to the center of the Earth, and it was judged that land acquisition delays would be many years in order to reach a machine radius to yield $\sqrt{s} = 90$ GeV.

The CLIC machine, on the other hand, is essentially always delivering events to the detector. So-called timestamping will be necessary to synchronize different parts of the detector so that correct events can be assembled offline. In other words, at CLIC the detectors should be as fast as possible, within the constraints of the physics requirements on resolutions.

5.3.1.1 Synchrotron Radiation

The behavior of electron machines is dominated by the radiation of energetic photons from the charged electrons as they are accelerated, and in a circular collider this acceleration is the familiar $a = v^2/R$. The radiation rate is the Larmor formula times the Lorentz boost factor to the fourth power, γ^4, or

$$P \text{ (energy radiated per second)} = \frac{e^2}{6\pi\epsilon_0}\frac{a^2}{c^3}\gamma^4 .$$

This can be cast in an easy form using the fine structure constant, $\alpha = e^2/(4\pi\epsilon_0\hbar c)$, $\hbar c \approx 0.2\,\text{GeV}\cdot\text{F} = 0.2\,\text{GeV}\cdot 10^{-18}\,\text{km}$, $\beta \approx 1$, and $\gamma = E/m$, as

$$P \text{ (energy/s)} = \frac{2}{3}(\alpha\hbar c)c\frac{\gamma^4}{R^2} .$$

A more useful number is the energy radiated per turn, and the time for one turn is $2\pi R/v$, so that

$$\Delta E \text{ (energy per turn)} = \frac{4\pi}{3}(\alpha\hbar c)\frac{\gamma^4}{R} .$$

For an electron machine, $\gamma = E/0.00051\,\text{GeV}$, and to keep the radiation rate low enough to avoid the necessity of powerful RF cavities to boost the energy back up, the ring should be made larger, the costs going into tunneling and civil engineering. This is the LEP solution (with an eye forward to the 7 + 7 TeV proton–proton LHC). For electrons, this is

$$\Delta E \text{ (GeV per turn)} \approx 8.85 \times 10^{-8}\left(\frac{E^4}{R}\right) .$$

E^4 is a brick wall, and LEPII encountered this limit at its highest operating energy ($E = 104.6\,\text{GeV}$, $R = 26.66\,\text{km}$):

$$\Delta E \text{ (GeV per turn)} = 0.4\,\text{GeV per turn} .$$

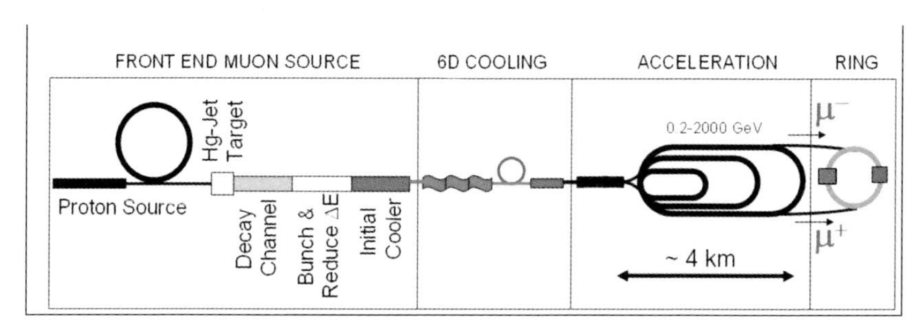

Figure 5.5 The configuration of a Muon Collider by main accelerator function.

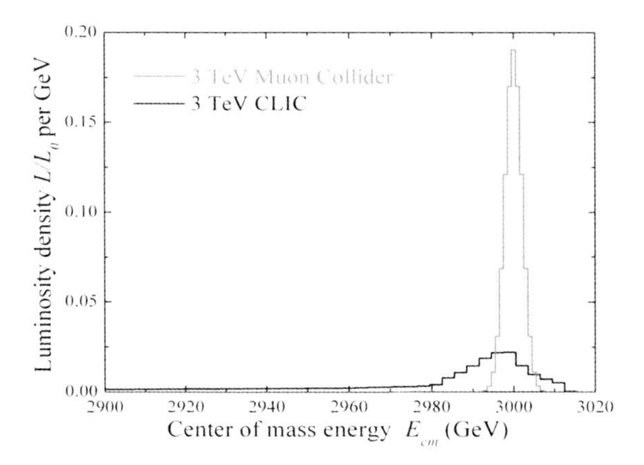

Figure 5.6 $\sqrt{s}$ at lepton–lepton collision.

For a Muon Collider on the Fermilab site, $R \sim 1\,\text{km}$ and $E \sim 3\,\text{TeV}$, the energy loss is $\Delta E \sim 5 \times 10^{-3}\,\text{GeV} \sim 5\,\text{MeV}$ per turn. In 1000 turns, the beams will scan across one beam width, Figure 5.6.

5.3.2
Proton Colliders: pp and $\bar{p}p$

The currently running hadron collider is the Fermilab $\bar{p}p$ Tevatron Collider operating at $\sqrt{s} = 1.96\,\text{TeV}$ that is anticipated to run through 2011. The next hadron collider has just been commissioned at CERN, the LHC, which sits in the LEP tunnel, and it will run at $3.5 + 3.5\,\text{TeV}$ for the next year. The RF frequency and bunch loading are such that the beams collide every 25 ns, a challenging but instrumentally possible time interval [90].

Apart from the beam crossing rate, a proton machine generally has a high radiation dose rate to the detectors, and therefore designers must be concerned with radiation damage to electronics and detector materials such as silicon strips and pixels, silicon chips, fibers, crystals, and scintillators.

The advantages of a proton machine are several: protons can be accelerated to higher energies, and there is no source limitation on the intensity of the beams, although the machine will limit the intensity through the beam–beam tune shift and the aperture. A physics advantage is that a proton machine is a "broadband" quark collider that can scan a wide mass region at once. This argument was once used to recommend against funding an e^+e^- machine.

It is remarkable that, historically, major discoveries have been made in equal measure on both proton and electron machines, in spite of their huge differences as far as detectors and collision energies are concerned.

5.3.3 Muon Colliders: $\mu^+\mu^-$

Accelerating unstable particles, in particular muons, has been proposed and studied [91–93], and now there are plans and active R&D at Fermilab, Figure 5.4, to produce, accelerate, and collide intense beams of μ^+ and μ^- at beam energies of up to 3 TeV. The incentive is to keep the physics advantages of an e^+e^- collider, that is, simple physics initial state, together with the machine advantages of a $p\bar{p}$ collider, that is, no radiation from the leptons. In contrast to Livingston's and Lawrence's afternoon of easy machine running, the design and building of a Muon Collider will push almost all the limits of machine physics at once. The downside to a Muon Collider is the simple fact that muons are unstable and decay into electrons while circulating in the beam. Therefor, the muon production, cooling, and acceleration must be done very rapidly, as illustrated in Figure 5.5.

For a muon with a rest frame lifetime $\tau_\mu = 2.2\,\mu\text{s}$ accelerated to 1 TeV, its lifetime in the collider frame is Lorentz dilated by γ:

$$\gamma\tau_\mu = \frac{E}{m_\mu} 2.2 \times 10^{-6}\,\text{s} \approx 22\,\text{ms}\,.$$

Therefore, an injector working at 10 Hz will fill the ring every 100 ms, and the beams will decay away to less than 1% before the next fill. However, in these 100 ms between fills, every muon in the beam will decay, and for 10^{12} injected, there will be 10^{12} decay electrons around the ring. It is a challenge to design both a machine lattice and a detector so that experiments are possible.[38] The beam-crossing interval is typically 6 to 25 μs, a very comfortable number for detectors.

5.3.4 Asymmetric Colliders: e^-p and B Factories

Some machines are fixed by their magnets to have equal energy beams, for example, LHC, LEP, and the Muon Collider. Other machines could collide at asymmetric energies, for example, ILC and CLIC, but there would be little physics motivation except as a means to suppress some backgrounds.[39] Three machines have been built deliberately to collide beams of different energies. The HERA ep collider studied deep inelastic scattering and searched for leptoquarks with 820-GeV protons and 27.5-GeV electrons, generally limited by machine considerations. The other two are the B factory machines[40] at SLAC and KEK in which asymmetric energy e^+e^- collisions boosted the final state $B\bar{B}$ system along the beam axis allowing for a lifetime separation of the B and $\bar{B}$ decay vertices.

38) A recent workshop at which some of these issues were addressed is the Muon Collider Physics Workshop, 10–12 November 2009, Fermilab, http://indico.fnal.gov/conferenceDisplay.py?confId=2855.

39) A. Mikhailichenko has suggested asymmetric e^+e^- ILC running to enhance the detectability of certain rare processes.

40) Proposed by Pier Oddone.

Why Not Bend Particles with E Instead of B?

The magnetic and electric forces on a moving charged particle are the *Lorentz* force

$$F_{\text{magnetic}} = e\boldsymbol{v} \times \boldsymbol{B} \quad \text{and} \quad F_{\text{electric}} = e\boldsymbol{E} .$$

In SI units, attainable stable laboratory fields are $E \sim 1\,\text{MV/m}$ and $B \sim 5\,\text{T}$, and a high energy particle has a velocity of $v \sim c = 3 \cdot 10^8\,\text{m/s}$, so that

$$\frac{F_{\text{magnetic}}}{F_{\text{electric}}} \sim v \cdot \frac{5\text{T}}{10^6\text{V/m}} \sim 1500 .$$

and thus magnetic deflections are far easier than electric deflections.

5.4 Problems

1. Do the synchrotron energy oscillations affect the center-of-mass energy of the collisions, $\sqrt{s}$, in a collider with RF cavities providing 100 MeV/turn?

2. A Muon Collider with 10^{12} muons per bunch at 1-TeV beam energy is circulating in the Tevatron, radius $r = 1\,\text{km}$. How many muon decays are there per second and per meter around the ring?

3. In a proton collider, the vacuum is 10^{-4} Pa.

   a) What is the beam-gas collision rate per meter?
   b) A more serious problem is very small angle Coulomb elastic scatters that nudge protons slightly outside the bunch into the beam "halo". Estimate the fraction of the beam that multiple scatters into the halo.
   c) A positive beam "pumps" the vacuum by driving N_2^+ and O_2^+ ions into the vacuum chamber walls. Make a model of this and estimate its effectiveness. It would not be possible to calculate an equilibrium pressure without assuming (or knowing) the out-gassing rate from the walls.

4. The acceleration rings for a Muon Collider can be "racetracks" as shown in Figure 5.5 or a single bidirectional linac connecting two circular turnaround rings at its ends. In either case the accelerating linacs are essentially made from the ILC 35 MV/m cavities, but the single linac geometry requires half the number of RF cavities in the linacs. (The bunch spacing and RF timing would be such as to allow acceleration in either direction for the correct bunches.)

a) For a turning radius of $R = 1\,\text{km}$ (just the Fermilab Tevatron radius), what is the synchrotron radiated energy loss per complete cycle (once around)?
b) For linac lengths of $L = 2\,\text{km}$ and using the ILC superconducting cavities with $E = 35\,\text{MV/m}$, how many times around the acceleration rungs is required to reach 3 TeV, and what is the total mean synchrotron energy loss?
c) What are the fluctuations in this energy loss?

6
General Principles of Big Detectors

"While SPEAR was being designed, we were also thinking about the kind of experimental apparatus that would be needed to carry out the physics. In the 1965 SPEAR proposal, we had described two different kinds of detectors: the first, a non-magnetic detector that would have looked only at particle multiplicities and angular distributions, with some rather crude particle-identification capability; the second, a magnetic detector that could add accurate momentum measurement to these other capabilities."

– Burton Richter (Nobel Lecture, December 1976)

We consider the gross features of a big detector: the configuration of the magnetic field, the tracking system, the calorimetry, and the extremities of the detector for the measurement of muons that penetrate the entire detector. The above remarks by Richter about the Magnetic Detector (as it was called before being renamed Mark I) and the following remarks by Jack Steinberger [94] about the ALEPH detector are interesting for their insight into the thinking that goes into a detector at the very beginning. This thinking is exactly the point of this book, although the careful reader may judge that I have not quite achieved this goal.

Steinberger and the Design of the ALEPH Detector There are very good detectors and there are excellent detectors, and in the latter category is the ALEPH experiment at LEP. Here let me quote Jack Steinberger [94] on the beginning design of ALEPH from nearly 30 years ago:

"We had open meetings about once a week, at which all important design features of what later was called the ALEPH detector were discussed and decided. We agreed to try to keep the design as simple and as uniform in technique as possible. The most important decisions were:

1. The magnet should be a superconducting solenoid with 1.5 Tesla (15 000 Gauss) magnetic field strength. Because of this high field strength and the required large size, 5.5 m in diameter, 7 m long, this represented a technical challenge.
2. The main tracking should be by means of a "time projection chamber" (TPC). The TPC is a very beautiful and conceptually simple device invented a few years earlier by David Nygren, an old friend and former collaborator, for the PEP e^+e^- collider at Stanford. It is a gas-filled cylinder in which the

Particle Physics Experiments at High Energy Colliders. John Hauptman

ISBN: 978-3-527-40825-2

electrons, liberated from the gas atoms by the passing charged particles, are drifted to the ends of the cylinder by a strong electric field. There they are detected with the help of proportional wire chambers, and their positions and arrival times are measured. This permits a three-dimensional reconstruction of the track, with a precision of about 0.1 mm in the transverse dimensions and about 0.7 mm along the drift dimension.

3. The electromagnetic calorimeter should be optimized for spatial rather than energy resolution, in line with its important role of particle identification. This fundamental insight came to us from Jacques Lefrancois, a French Canadian and professor at Orsay near Paris. It should be inside the magnet in order not to suffer from the degradation of the particle by their interaction in the magnetic coils and tank. The result was a 45-layer sandwich of 3-mm lead sheets separated by 5-mm-thick wire chambers, capacitively coupled to square pads, about 3 cm × 3 cm, arranged in 75 000 towers, projecting to the collision point. The energy deposited along the way in the towers was sampled in three stories.
4. The hadron calorimeter should use the iron return yoke of the magnetic flux, which, conveniently, needed to be of similar thickness as that required for the hadron shower development. The iron was divided into 24 layers, each 5 cm thick, the layers were separated by 2 cm gaps, which accommodated the simplest and most economical detectors known, so-called "streamer" wire chambers, which sensed the number of traversing particles. These could be read out as wire planes and, capacitatively, in 6500 projecting towers.
5. The detector naturally consisted of a central "barrel" and two "end caps". It was agreed that the calorimeter technologies should be the same in the barrel and the end caps, and that the two detection planes surrounding the whole detector, signaling that muons that had traversed the hadron calorimeter, would be of the same streamer chamber design as the hadron calorimeter detectors.

This detector was proposed in January 1982 and approved by the LEP committee in July 1982.

There are several lessons here. First, it is much easier for laboratory physicists to meet regularly (weekly, in this case) than it is for university physicists to travel, coordinate, and meet regularly to talk through a design. Second, the uniformity of technologies is consistent with item (4) of the principles listed in Section 6.2, and it runs counter to the funding and political imperatives referred to in Section 8.5. Third, an electromagnetic calorimeter with three depth sections was common at that time, and it was a "projective geometry" calorimeter. Fourth, for reasons of economy the hadronic calorimeter was built into the magnetic iron flux return yoke. Finally, the main challenge noted as item (8) of Section 6.2 is to see these lofty principles through to the completion of a functioning detector.

6.1 Detectors at Big Colliders

Only the LHC has achieved reality in recent years, but each of the big colliders (SSC, ILC, CLIC) had at least the beginnings of detector groups and detector designs. A group proposing a big experiment always writes a "design document" and eventually a technical design report (TDR), but the real question is what happens before this point and how were the decisions made. It is my opinion that some of these design decisions are driven by what-I-did-on-my-last-experiment or what-my-lab-can-build considerations, properly upgraded and enhanced for the new machine. We might note that Steinberger's group did exactly the opposite: design with the newest chambers (TPC and streamer chambers), innovate with the electromagnetic calorimeter in both physics purpose and projective geometry, and, most importantly, build a large solenoid so that the physics instruments inside it have enough depth to make good measurements.

There is usually severe time pressure to design a detector and to present it to the laboratory program advisory committee and, for components that are still under R&D studies, to present various options in the event of R&D failures.

For the ILC, several detectors [95] have been under design, study, and testing for several years. This is unusual, but novel and a good idea. Barry Barish leads the Global Design Effort (GDE) for the ILC under which the detectors are being designed as "concept detectors" for which rather high instrumental requirements [96] have been defined that the detectors must meet, *viz.*

1. Momentum resolution at high momenta:

$$\frac{\sigma_p}{p^2} \approx 5 \times 10^{-5}\,(\mathrm{GeV/c})^{-1}\,.$$

This requirement allows measurement of the ZZH coupling in the process $e^+e^- \to Z^0H^0 \to \mu^+\mu^-X$ by a precision momentum measurement of the $\mu^+\mu^-$ system, and therefore a precise missing mass measurement against the Higgs. This momentum resolution is required to achieve a H^0 mass resolution of $\sigma_H \approx 1\,\mathrm{GeV/c^2}$ in the $120\,\mathrm{GeV/c^2}$ Higgs mass region.

2. Energy resolution on jets:

$$\frac{\sigma_E}{E} \approx \frac{30\%}{\sqrt{E}}\,.$$

This requirement allows direct separation of $W^\pm \to jj$ from $Z^0 \to jj$ by two-jet mass resolution, and also two-jet background rejection against the W and Z.

3. B-tagging by impact parameter: a goal of 80% b-tagging efficiency at $\epsilon_b \approx$ 90% b efficiency allows the reconstruction of the four bs in final states such as $e^+e^- \to Z^0H^-H^0$ and $e^+e^- \to t\bar{t}H^0$, and to separate these from the huge non-b and two-b backgrounds. Purity and efficiency are both important since the overall event efficiency is proportional to ϵ_b^4.

These dramatic instrumentation requirements can be loosely characterized as "two to ten times better" than the already excellent LEP detectors, OPAL, L3, DELPHI, ALEPH. Since no detector can presently achieve these specifications, more time is very useful. There are further requirements on the machine and the beams.

6.2
Design Principles

First, we will list a set of eight design principles that were the guide for the design of the 4th detector at an electron linear collider (Chapter 7) and may serve here as a challenge for thinking about larger issues.

1. *Independence of each detector subsystem*: each major detector subsystem should be able to "stand on its own two feet", for example, the calorimeter should function as a device to measure energies absorbed in specific volumes without dependencies on other subsystems such as tracking. Similarly, the tracking system should measure and recognize patterns of contiguous spatial measurements as individual tracks without depending on the calorimeter or vertex chamber. A muon system is always at the extremity of the detector and should be able to identify and measure muons independently of the tracking system and the calorimeter that precedes it, including muon traversal of many nuclear interaction lengths of material. It is always true in the physics analysis of events in a detector that all systems will be used together to maximum benefit, for example, a high precision vertex chamber will assist pattern recognition in an outer tracker, the tracker and calorimeter together will discriminate between $e^{\pm}$ and γ clusters in the calorimeter, and the muon system will depend on the momentum measurement in the track and the absorption of hadrons in the calorimeter.
   Justification for this *independence* principle is partly that it avoids correlated weaknesses in the detector, that is, if the calorimeter depends on a certain level of performance in the tracking system (efficiencies, fake track rates, etc.), then a decrease in one or more of these will decrease the performance of the calorimeter, and there will be two weak detector systems instead of one. A second reason is that it is hard enough to design and build an excellent tracker or an excellent calorimeter, but it is even more difficult to make sure that one serves the other in specific ways.
2. *Particle identification* is critical to all physics at a collider, for particles at both high and low momenta, and as many standard model partons as possible must be identified by direct measurements in independent detectors.
3. Each independent detector must be of sufficient capability that *auxiliary or ancillary detectors are unnecessary*. Examples of auxiliary devices are tail catchers for a too-shallow calorimeter, preshower devices to tag showers developing in too-massive detector materials before reaching the calorimeter, additional multiple

tracking devices to assist either the momentum resolution or pattern recognition in a central tracker, end-cap chambers to remeasure tracks after too-massive end plates in the central tracking system, interdetector chambers to compensate for dead volumes, or multiple different technologies for reasons of measurement redundancy.

4. *Use common technologies* wherever possible for reduced costs, ease of construction, ease of calibration, and general uniformity in the response of the detector. This runs directly counter to the sometimes political needs of a large collaboration to secure funding from many sources but can be enforced at the beginning (see Steinberger comments above).
5. *The magnetic field configuration is a big decision*: the worldwide favorite is a solenoid providing a uniform B_z field for precision tracking. We will discuss the merits of an iron-free detector such as that introduced and designed for the 4th detector. The problems of fringe fields, especially in the delicate machine-detector interface region of future precision colliders, the problems of internal forces on magnetically permeable volumes, detector movement in the interaction region (IR), ground settling in the IR, and intradetector spatial surveying requirements will be discussed. Some of these problems can be solved with more iron in the poleface regions and elsewhere, but *more iron means less detector*.
6. The *time domain on the nanosecond scale is powerful information for detectors* and is often overlooked, sometimes for good reasons such as cost.
7. *Relative absence of inactive, or dead, regions within the instrumented volume*. This is unavoidable for the support of massive instruments such as the calorimeter and the superconducting coils, but it is quite avoidable in calorimeters in which all readouts can be at the rear, and for lightweight tracking systems that do not require cooling or fragile supports within the tracking volume itself. At future higher energy colliders, it will be essential to measure track trajectories with very high spatial precision, and this implies silicon[41] with microetching at the 5- to 20-μm level, which implies in excess of 10^9 channels of electronics with its accompanying heat load, and therefore cooling that cannot be achieved without extra mass. This will be the main problem in future tracking systems in terms of keeping the material budget as low as possible.
8. A final strength of a big detector (especially one built under scheduling or funding duress) is a design that is *resistant to "engineering creep"* (for lack of a better term). All big detectors undergo the successive transitions from ideas, to physics prototypes, to beam tests, to engineering designs, to larger-scale prototypes, to system beam testing, to a large-scale industrial manufacturing stage, and, finally, to installation and *in situ* testing. During these stages of a big project, engineering necessities of tolerances, gravitational supports, and

41) Many groups are developing ingenious silicon sensors, and organized research groups are pursuing the development of whole tracking systems based on silicon, such as the SiLC (Silicon for the Linear Collider) collaboration led by A. Savoy-Navarro (http://lpnhe-lc.in2p3.fr/).

internal standoffs and supports are almost always met by adding materials to the detector.[42] Good physics intentions are lost at this point.

In the simplest sense, a good detector should have a zero-mass tracking system (vertex and central tracker) inside a high magnetic field that is perpendicular to all tracks, and an infinite-mass calorimeter that only muons (and neutrinos) can penetrate, all of which are continuously sensitive on the nanosecond time scale and which have momentum and energy resolutions for two-body decays that are comparable to the intrinsic widths of important states, for example, $\Gamma_W \sim \Gamma_Z \sim 2\,\text{GeV/c}^2$.

6.3
Magnetic Field Geometries

Most collider detectors both large and small have followed Richter's successful detector design for Mark I.

The first consideration is a detector with a magnetic or without a magnetic field, and both have been prominently and successfully used,[43] but the universal choice is a strong magnetic field in the multi-Tesla range. The next consideration is the geometry of the magnetic field: solenoidal, toroidal, longitudinal, transverse, or "split field" (SFM), and most have been tried or at least seriously studied [97]. The famous Mark I detector had a 0.4-T solenoidal field, but a competing early design was nonmagnetic [98]. The design goal is to present a $\boldsymbol{B}$ field that is perpendicular, or nearly perpendicular, to the directions of particles everywhere, or almost everywhere, in order to take full advantage of the sin-term in the Lorentz force, $\boldsymbol{F} = e\boldsymbol{v} \times \boldsymbol{B}$, for momentum measurement.

Solenoid The predominant choice for big colliders today is a solenoidal magnetic field, which is easily calculated as $B = \mu_0 n I$, for n = number of turns per meter with current I. This field is rather uniform on the interior of the solenoid, falling off at the ends where the axial component B_z is about one-half its value at the center. The advantages of a solenoid are that (i) a uniform field can be relatively easily achieved with Cu, Al, or superconducting coils, (ii) axial wire chambers can be put inside the solenoid, including a TPC that requires a large volume uniform $\boldsymbol{B}$ field, (iii) geometric track reconstruction is easier since tracks are close to perfect helices describable with only five parameters, and (iv) geometrical acceptance is high over a large fraction of the solid angle. The only disadvantage is that tracks in

42) Both the CMS and ATLAS detectors have more than one radiation length (X_0) of material in their tracking systems near $\eta \approx 1.5$ (always a difficult region for matching the barrel to the end cap in a solenoidal geometry). For events with jets and many $\pi^0 \to \gamma\gamma$ decays per jet, there will be many electromagnetic showers starting inside the tracking system. This was not intended in the physics design.

43) Most recently, the D0 experiment during run I of the Tevatron $\bar{p}p$ Collider at 1.8 TeV (1992–1996) codiscovered the t quark along with CDF (with a solenoidal magnetic field); however, a 2-T solenoid was installed for run II.

Figure 6.1 Photograph of the CMS solenoid inside its cryostat and inside the muon system iron yoke [100].

the forward region with $\theta < 10$–$20°$ have poor momentum resolution due to the small component of the particle momentum that is perpendicular to the field. Since this region may also be obscured by particle debris from the beams, sometimes an experimenter chooses to not even instrument this region, but this decision is very machine dependent. However, for a future high precision collider such as the ILC, it is planned to instrument the entire forward region with rad-hard calorimeters down to $\theta = 3.9$ mrad [99].

It should be remarked that "loopers", those particles with a gyro diameter less than the radius of the solenoid, will exit the tracking volume axially and will not remain trapped and spiral through the tracking chambers. Experiments with the same basic geometry as the Mark I are HRS, MAC, TPC, TASSO, Pluto, ALEPH, DELPHI, L3, OPAL, CDF, DO-II, CMS, VENUS, TOPAZ, Belle, BaBar, Mark II, Mark III, AMY, CLEO, BESS, ZEUS, H1, SLD, and ATLAS.

The large solenoid of the CMS experiment is shown in Figure 6.1 with $B = 4$ T. It should be emphasized that the risk with the (superconducting) solenoid is total: even a partial failure will stop the experiment from running until a repair or a replacement is made [100]. This is not necessarily true for other systems since you can lose the tracking system, or a part of it, and still do physics with a fraction of the solid angle.

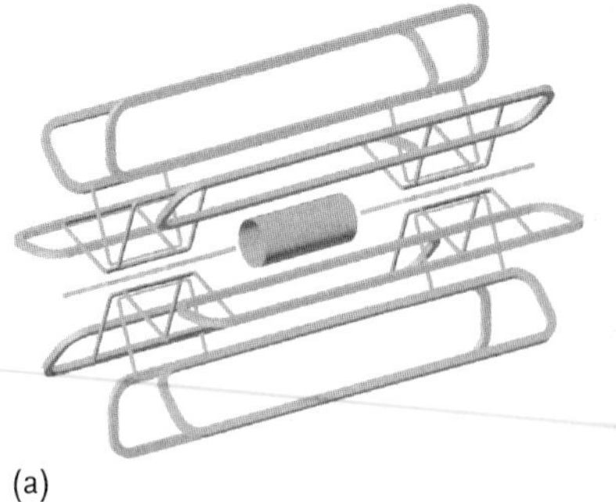
(a)

(b)

Figure 6.2 (a) Conductors of the ATLAS experiment: a central solenoid for tracking in a silicon system, surrounded radially by a toroid consisting of six conducting loops, and bounded axially by two end-cap toroids each with six conducting loops; and (b) a photograph of the toroid cryo enclosures during construction of the ATLAS detector.

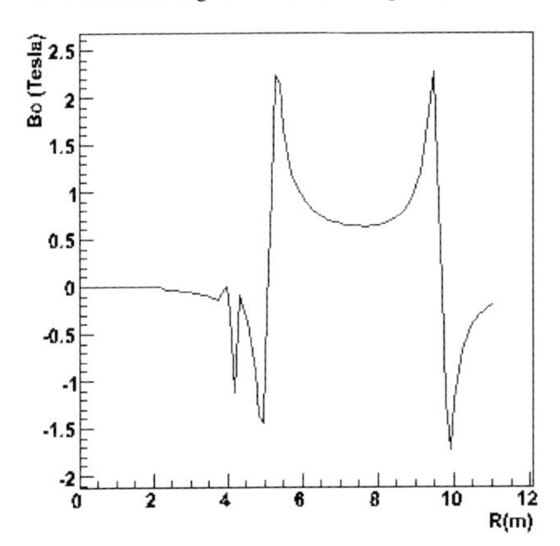

Figure 6.3 A calculation of the $B_\varphi(r)$ field through the barrel toroid.

Toroidal One major exception to a solenoidal field is the toroidal field of the ATLAS experiment at the CERN LHC generated by discrete conductors elongated in the axial (beam) direction, Figure 6.2. The inner radius of the central toroid is 5 m and the outer radius 10 m.[44] In spite of this huge size, the internal forces of the detector are much smaller than with an iron flux return yoke. The magnetic field is not uniform, and tracks near the conductors will pass through a highly complicated B_φ field that varies strongly in radius, r, as shown in Figure 6.3.

The field lines in a cross-section at the center of the detector ($\theta \approx 90°$) are shown in Figure 6.4a and in a cross-section through an end-cap toroid ($\theta \approx 10$–$20°$) in Figure 6.4b. The toroid field must scale like $1/r$ by Ampere's law, and the field intensity plot shows the low-radius end-cap toroid with $B \geq 1$ T, whereas the central toroid field is $B \sim 0.3$–0.5 T.

Therefore, for tracks leaving the interaction point (IP) that are mostly radial, p_r, passing through the central toroid with a field that is mostly azimuthal, B_φ, the

44) A similar design was discussed by Larry Jones at the Berkeley SSC Workshop (1987) with combined barrel and end-cap toroids.

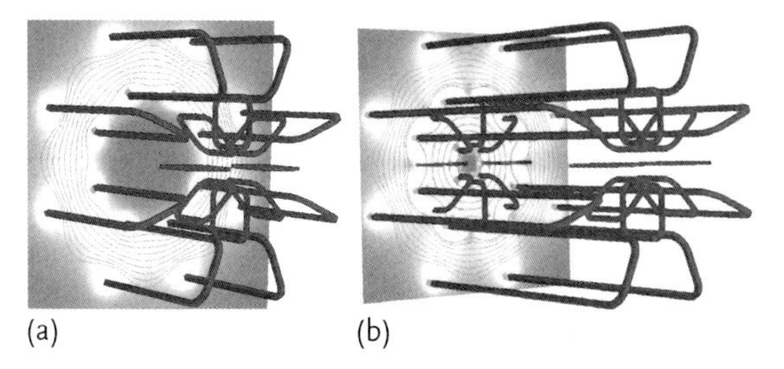

Figure 6.4 The azimuthal magnetic field of the ATLAS toroids for a slice through (a) the center ($z = 0$) of the detector and (b) through the end-cap toroids near $z = \pm 10$ m.

Lorentz force is axial and the particles are bent toward $\pm z$. For tracks at small angles, p_z, passing through the end-cap toroids with a mostly azimuthal field, B_φ, the force is radial and the particles are bent either out or in, $\pm r$. This field configuration has achieved what the solenoid has not: complete coverage of the solid angle for charged tracks. The price in the ATLAS design is a magnetic field so complicated that it is not uniform in r, θ, or ϕ.

Transverse A transverse $\boldsymbol{B}$ field is a uniform field (like a window frame, or C frame, magnet) in the beam providing a vertical field, B_z, around the interaction point.

Longitudinal This field could be generated by two Fe pole tips upstream and downstream of the interaction point, with holes for the entry and exit of the beams. Locally, it would be like a solenoidal field, but highly nonuniform.

Split Field A major experiment at the CERN pp Intersecting Ring collider (ISR) was the Split Field Magnet (SFM) with a $\boldsymbol{B}$ field pointing up on one side of the IP and pointing down on the other side, with zero field exactly at the IP. The circulating beams were bent in one direction, then bent back. The scientific driver was studies of particle production in the forward direction, so-called diffraction or "Pomeron" exchange reactions, and the uniform field presented to the forward-going particles provided excellent momentum bending. Of course, like the "transverse" field, particles traveling vertically parallel to the $\boldsymbol{B}$ field had poor momentum resolution.

Dual Solenoids The newest addition to detector field geometries is the dual solenoid, shown in Figure 6.5, and advocated in the 4th concept detector, Chapter 7. The currents of the inner solenoid are driven in one direction, and the currents in the outer solenoid and the end coils are driven in the opposite direction. Therefore, the field generated by the inner solenoid is redirected radially outward by the end coils and then directed axially in the opposite direction by the currents of the outer solenoid. The dimensions in this figure are nominal, but almost all radii and axial extends will produce a decent field. With some care [101], the currents can be tailored to produce a highly uniform field for the tracking system inside

$r, z < 1.5$ m, a uniform field in the annulus between the solenoids, and an almost exactly zero fringe field.[45] In Figure 6.5, the field at the ends of the inner solenoid is mostly radial, and the forces on the azimuthal conductors is axial so as to compress the solenoid. These forces can be large, and a more sophisticated design [101] of the ends of the inner solenoid has alleviated these forces and also limited the maximum field on the superconducting cables to about 5 T.

I believe there are many advantages to the dual-solenoid field configuration, and these are presented in Sections 7.1.4–7.2. They range from physics advantages (better muon physics) to machine advantages (no fringe field) in the delicate region where the final focus magnetic elements are located.

The disadvantages are that the outer solenoid is large and the iron that is now missing cannot serve as a radiation shield. The first problem may be solvable in that the outer solenoid is only large; it is not necessarily a precision solenoid since it must (mostly) just channel magnetic flux. Radiation protection depends on the IR (will there be a second detector in the same IR?) and the machine (is it a proton or electron machine?), and this is a very complicated business requiring experts [18, 19].

The CMS-style conductors are shown in Figure 7.14 for a relatively compact solenoid. The field quality is excellent, as shown in Figure 6.7, in which each color gradation is $\Delta B/B \approx 6 \cdot 10^{-5}$, leading to 10^{-3} variation inside ± 1.5 m and $< 1.4 \cdot 10^{-4}$ inside ± 1.2 m. The mapping of the field will be done with NMR probes throughout the entire volume of the detector with absolute accuracy of $< 10^{-4}$. The field uniformity is shown for slightly elongated solenoids in Figure 6.8.

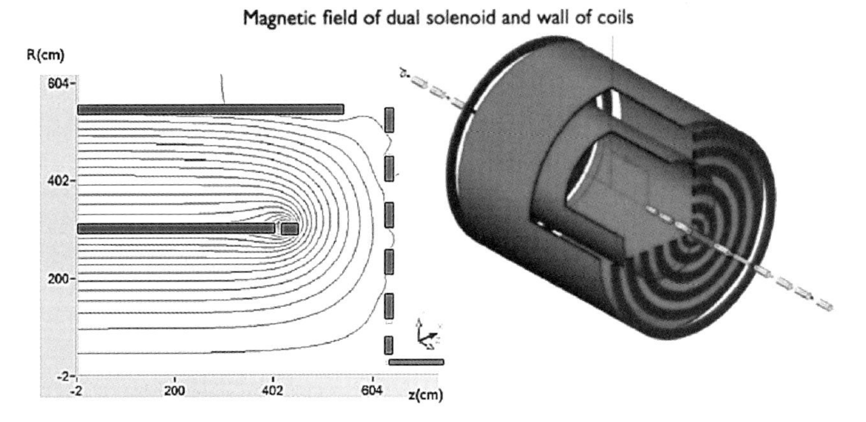

Figure 6.5 The magnetic field configuration of dual solenoids. The central tracking field is $B_z = 3.5$ T, the reverse field between the solenoids is $B_z = -1.5$ T, and the field is symmetric in ϕ and smooth everywhere in θ. The field can be exactly reversed to cancel detector asymmetries in high precision quark asymmetry measurements, and the fringe field is essentially zero. The main problem is the construction of the larger solenoid.

45) We have not deliberately designed for a zero fringe field, but it is always small and, if it were nonzero, one strategy would be to map its multiple components, then set up appropriate conductors to exactly cancel these multiple components.

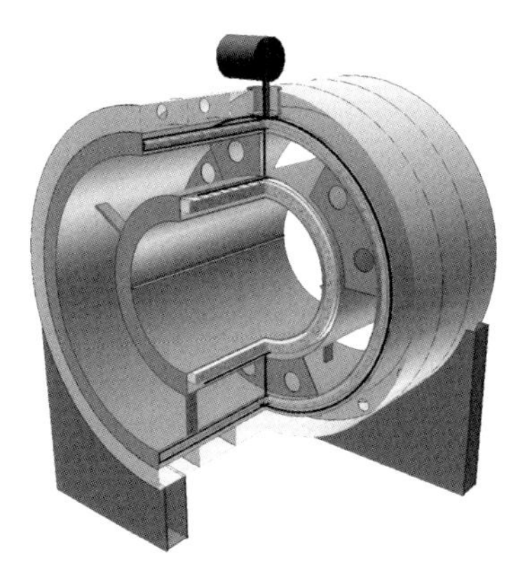

Figure 6.6 Cut-away of the dual solenoids with supports.

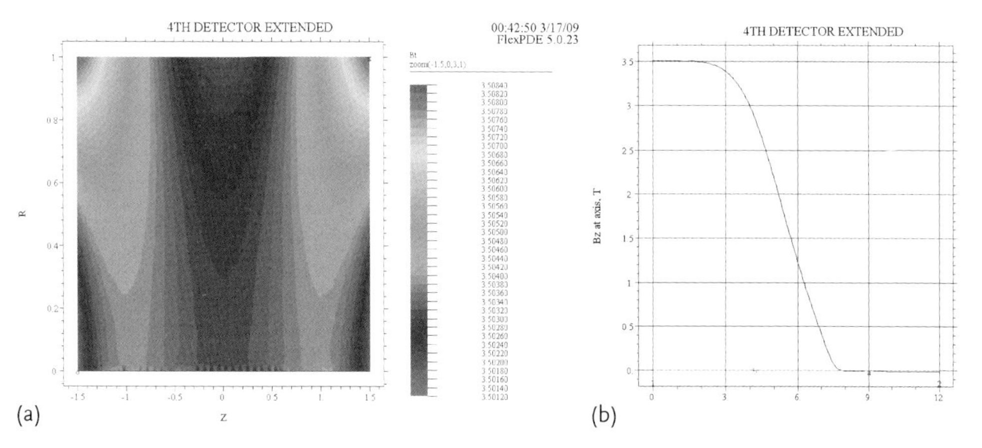

Figure 6.7 The B field quality is excellent: (a) each color gradation is $\Delta B/B \approx 6 \cdot 10^{-5}$, and (b) the main axial field B_z goes to zero at 7.5 m.

This design has safe peak $\boldsymbol{B}$ fields inside the superconducting cables, as shown in a blowup of the region of the Helmholtz coils at the ends of the inner solenoid in Figure 7.16. A view of the dual solenoids in their cyrostats in the IR is shown in Figure 6.6.

The magnetic field in this air volume between the two solenoids will back-bend the muons for a second momentum measurement (after the calorimeter) to achieve high precision without the limitation of multiple scattering in Fe. This field is filled with the same technology CluCou wires (within tubes, as in ATLAS) and will be read out by cluster timing electronics and therefore also serve as a continuous volume time monitor of all activity outside the calorimeter. If anomalously ionizing particles appear outside the calorimeter, the 3.5% dE/dx-specific ionization measurement by cluster counting may provide some particle identification for these particles.

The outer solenoid field will subtract from the inner solenoid field. For currents and turn densities of i_1, n_1 for the inner solenoid and i_2, n_2 for the outer solenoid,

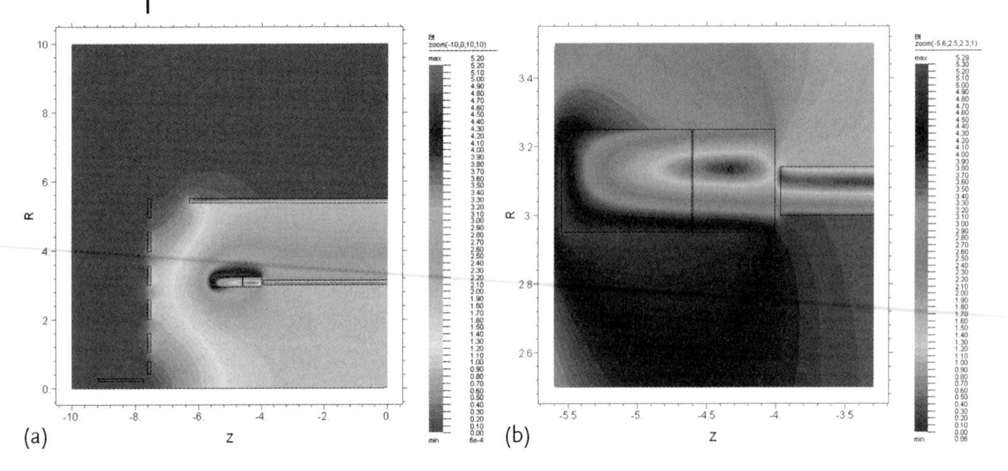

Figure 6.8 (a) The spatial uniformity of the B field is excellent, and very acceptable for the tracking system. The fringe field is essentially zero. (b) A detail of the field intensity in the Helmholtz ends showing a completely acceptable peak field of $B = 5.08\,\mathrm{T}$.

the field in the center for tracking will be approximately

$$B_{\text{tracking}} \approx \mu_0 n_1 i_1 + \mu_0 n_2 i_2 \,.$$

For the 4th design, Section 7.1.4, these numbers are $\mu_0 n_1 i_1 \approx 5\,\mathrm{T}$ and $\mu_0 n_2 i_2 \approx -1.5\,\mathrm{T}$, so that the central tracking field is about 3.5 T. The annulus region between the solenoids is a tracking region filled with low-mass tracking chambers of any convenient kind, or for speculative searches it can be filled with a high precision time-of-flight system that is continuously sensitive, or another calorimeter system, or a system capable of seeing something odd, for example, the byproducts of an LSP[46] with a finite lifetime that might wander out away from the IP and decay after many meters.

For an ordinary momentum measurement as a spectrometer in the annulus ($r_1 \to r_2$) between the solenoids, the momentum resolution does not depend on the radius (r_2) of the outer solenoid. This is because the number of Webers in the annulus is the same as the number of Webers in the inner solenoid since all of the flux of the inner solenoid is returned through the annulus. If B_1 is the field in the center in Tesla (T = Weber/m^2) and the area is πr_1^2, then the field B_2 in the annulus is given by

$$\mathrm{W\ (Weber)} = B_1 \pi r_1^2 \approx B_2(\pi r_2^2 - \pi r_1^2) \approx B_2 \pi r_2^2$$

since $r_2^2 \gg r_1^2$, and so $B_2 \approx B_1 (r_1/r_2)^2$. The momentum resolution by a sagitta measurement in the annulus will have a resolution that scales as $1/B_2$ and $1/(r_2 - r_1)^2$ since $(r_2 - r_1)$ is the track length in the field. The second solenoid can be larger so that σ_x is not important, although calibration over long distances is very difficult.

46) Lightest SUSY Partner, usually assumed to be absolutely stable.

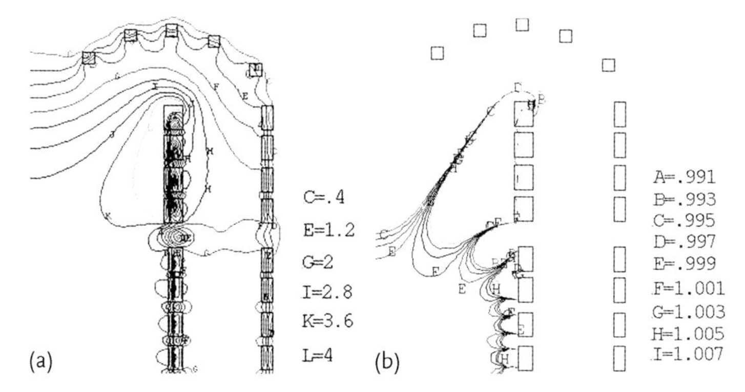

Figure 6.9 (a) Dual solenoid with curved end-wall coils showing a maximum field intensity of only 5.003 T, lower than the CMS-like 5-T solenoid design. (b) The dual solenoid with curved end-wall coils, and still excellent field uniformity in the tracking region. Each color gradation is $\Delta B/B = 0.1\%$.

Alternative dual-solenoid designs exist, for example, as shown in Figure 6.9a,b by Wake [102].The designs by Wake are novel and interesting and display an array of possibilities for detectors that deserve scrutiny. It is my feeling that almost any dual-solenoid configuration between the fields of Figures 6.5 and 6.8 and that of Figure 6.9 are possible.

Force on Permeable Material Inside B field

As the iron pole tip of area πR^2 moves a distance dz into the field region, Figure 6.10, the energy density of the magnetic field in the volume element $\pi R^2 \cdot dz$ decreases from $B^2/2\mu_0$ to $B^2/2\mu$. This is a huge decrease in energy density by a factor of $(\mu/\mu_0) \approx 10^3$ that is stored in the magnetic field, or

$$du(\mathrm{J/m^3}) = \frac{B^2}{2\mu} - \frac{B^2}{2\mu_0} \approx \frac{B^2}{2\mu_0} \, .$$

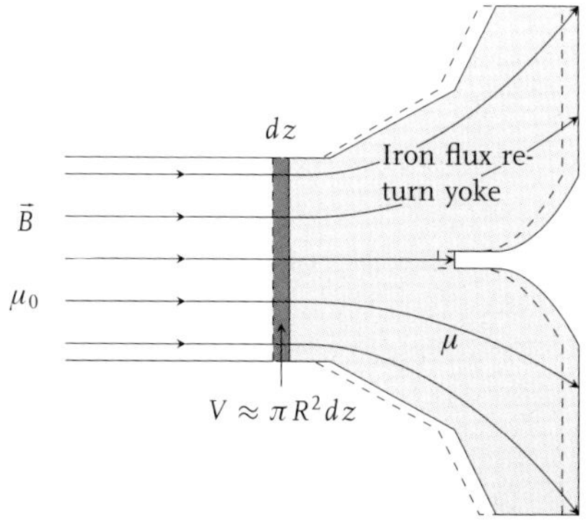

dz
Iron flux return yoke
$\vec{B}$
μ_0
μ
$V \approx \pi R^2 dz$

Figure 6.10 Force on permeable material in a magnetic field.

The total change in the magnetic energy of the system is

$$d\,U(\mathrm{J}) \approx \frac{1}{2\mu_0} B^2 \cdot \pi R^2 \cdot dz\,,$$

and the corresponding force is

$$F_z = -\frac{d\,U}{dz} = \frac{B^2 \pi R^2}{2\mu_0}\,.$$

For $B = 4\,\mathrm{T}$ and $R = 1.5\,\mathrm{m}$, the force is $F_z \approx 12\,000\,t$, and the iron is pulled toward the region of higher B^2.

The forces on magnetically permeable materials inside magnetic fields can be very large and distort the geometry of the detector.[47] Such spatial distortions may not be a concern if spatial precision is not important for particle measurements, for example, gross jet measurements in the CMS forward (HF) detector that, in fact, moved axially. Furthermore, the movements can be measured and easily corrected in the software and reconstruction codes, although due to hysteresis effects the movements are not always reproducible, but, again, these can be calculated. However, for the next generation of precision detectors at future lepton colliders in which spatial precision requirements will be a factor of ten more demanding, a detector without these forces may be desirable. It might be added that the forces on the inner solenoid of the dual-solenoid geometry are small and stable, that is, a small displacement from the center increases the magnetic stored energy and the restoring force is both toward the center and small.[48]

The idea of noniron flux return and using additional conductors to eliminate the iron mass has expanded [103] to include a hybrid of a conventional iron flux return with end coils to redirect the axial flux out radially, and in this way to reduce the volume of iron required in the end-cap region, as shown in Figure 6.11a and b. The superconducting solenoid is near $r \approx 4\,\mathrm{m}$ and extends axially to nearly $z \approx 4\,\mathrm{m}$. The iron volumes are shown around the solenoid; the arrow in the lower right in Figure 6.11b points to a compensation solenoid for pretwisting the ribbon beams (at 14 mrad crossing angle) before they enter the solenoidal field, which twists them back to horizontal. A conventional magnetic yoke for flux return of a solenoid is shown in Figure 6.11a, while Figure 6.11b is a magnetic field configuration that is mostly an iron flux return yoke, assisted by end coils that redirect the axial flux outward radially, much like the end coils in the dual-solenoid configuration, Section 6.3, in the 4th concept detector, Chapter 7. The beneficial results of

47) See the recent article "Motions of CMS detector structures due to the magnetic field forces as observed by the Link alignment system during the test of the 4 T magnet solenoid", L.A. García-Moral *et al.*, *Nucl. Instr. Meths.* **A606** (2009) 344.

48) Calculations by B. Wands and A. Mikhailichenko, http://www.4thconcept.org/4LoI.pdf (appendices).

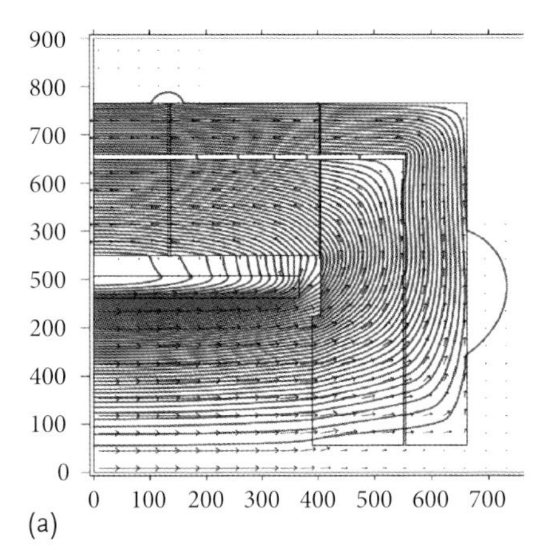

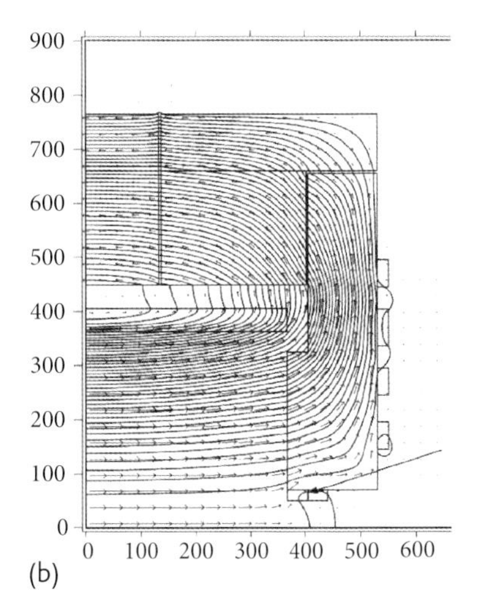

Figure 6.11 A recent configuration in the evolution of the ILD (International Large Detector) magnetic field configuration using end coils as in the 4th concept.

this hybrid design are many: the iron is maintained as superstructure for support and for radiation shielding, there are two more meters of axial space in the IR, and several other problems in the final focus involving the final quadrupole (QD0) are easier. This will have many benefits for final focus optics, machine-detector interface, more available axial space in the IR, and a more effective final quadrupole (QD0 in the ILC nomenclature) positioning.

There are many degrees of freedom in dual-solenoid geometries, for example, designs by M. Wake [102] (KEK laboratory) of several dual-solenoid geometries, all of which could work for an experiment, with end coils that match the field and lumped coils to achieve both a uniform tracking field and acceptably low fields at the conductors, as shown in Figure 6.9. The tracking field region (3.0 m axial and 2.4 m radial) is uniform to 0.1%, the maximum current density in the coil is

18 kA/m^2, the peak field is 5.0 T, the total system stored energy is 5.08 GJ, and the total inductance is 25.4 H. The main message is that completely acceptable magnetic fields can be achieved with dual solenoids in big experiments, and that there is likely a continuum of interesting solutions between and beyond this configuration and the configuration of the 4th design in Figure 6.8.

The stored energy, U, in a solenoid is the integral of B^2 over all space,

$$U = \frac{1}{2\mu_0} \int B^2 dV \,, \tag{6.1}$$

so a thin superconducting solenoid has $U = (B^2/2\mu_0)\pi R^2 L$, while the energy stored in the iron mass is negligible. The momentum resolution goes as $\sigma_p/p \approx p/B\ell^2$ and ℓ is about equal to the radius, R. However, a prudent solenoid design would limit U/M to a reasonably safe value, like 5–7 kJ/kg, which by inspection is the only "constant" in Table 6.1. The $U/M = 12$ kJ/kg of CMS was a risky extrapolation beyond all previous solenoids.[49]

6.3.1 Summary of Magnetic Field Geometries

Four main variations on the configuration of the magnetic field are shown here. The most common is the solenoidal field B_z, Figure 6.12a, with the interaction point at the center and tracking chambers arranged as cylinders inside the solenoid. The main strengths are that the field is spatially uniform throughout the tracking volume, the field intensity can be large (from 0.1 T with Cu windings up to 4 T with superconducting windings), high transverse momentum particles (or particles perpendicular to the beam) are perpendicular to the field and therefore their momentum is well measured, the paths of the particles are unobstructed by coils, and a solenoid is relatively easy to construct. The weakness of a solenoid is that forward-going tracks (polar angle $\theta \lesssim 20°$) are poorly measured since only a fraction of the momentum vector is perpendicular to the field and, in a realistic tracking chamber, there are fewer radial measurements. This problem is alleviated by placing disk chambers are low angles, but nothing will help as $\sin\theta \rightarrow 0$.

A toroidal field B_ϕ, Figure 6.12b, such as in the ATLAS detector, can be generated to surround most of the solid angle so that radial tracks (p_r) are bent axially. The end regions can also be equipped with toroids (as in ATLAS) so that axial tracks (p_z) are bent radially. At $B_\phi \sim 1$ T and a sufficient arc length ℓ, good momentum resolution can be obtained over nearly the whole solid angle, and without the problems of iron. The problems are that the solid angle is obscured by the coils that generate the field and, since the field a distance s from a conductor goes as $B \sim 1/s$, the field is nonuniform, as seen in Figure 6.3. This is not a problem if the field is known well.

49) Alain Herve, private communication. The cool-down of the CMS solenoid took 1 month, and it was 100% critical for the success of the experiment: no solenoid, no experiment. This is one of the largest risks in a big experiment.

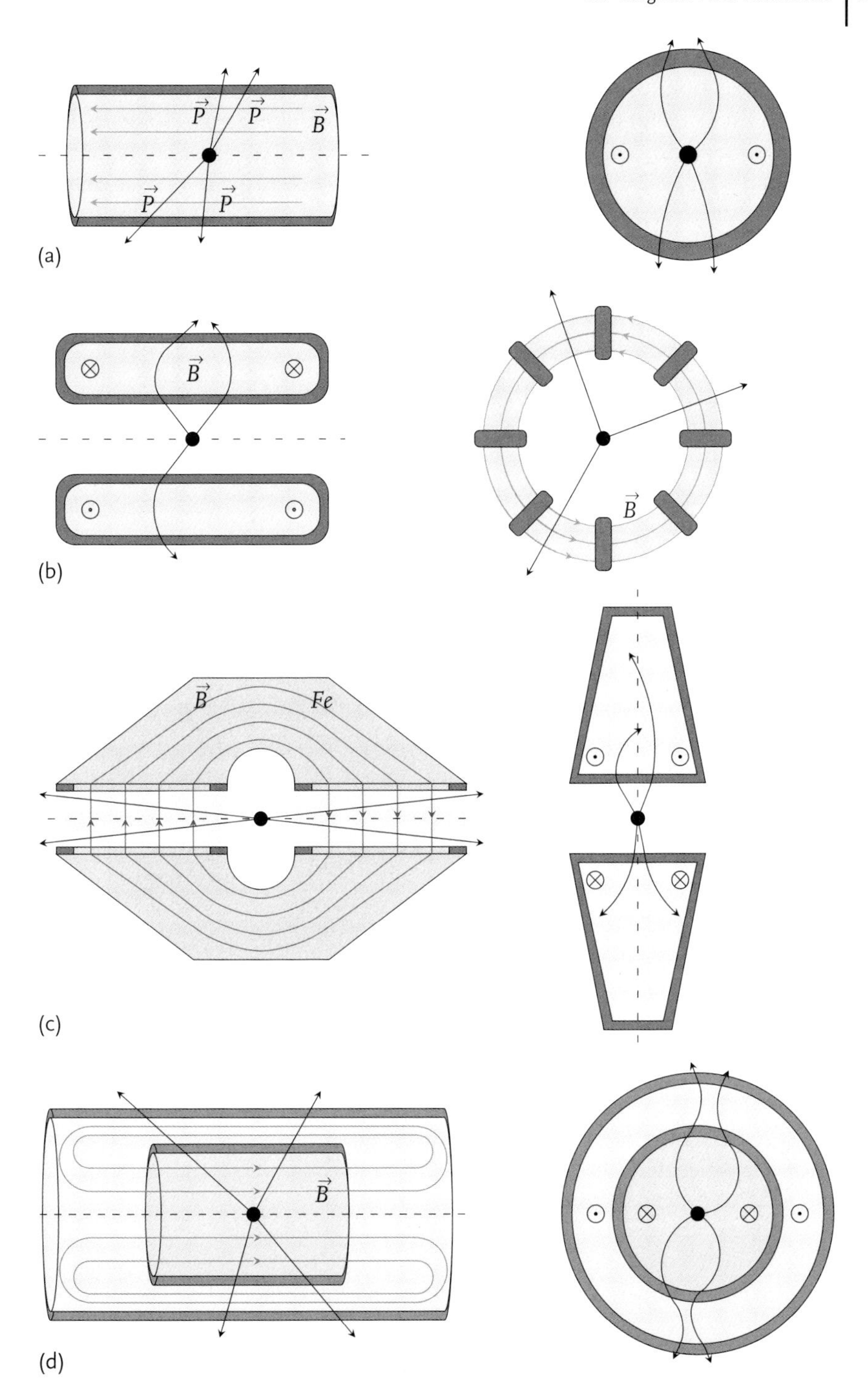

Figure 6.12 Four different magnetic field geometries that have been used in big collider experiments. The lines show the direction of charged tracks in these fields. (a) solenoidal field B_z; (b) toroidal field; (c) split field magnet experiment at the CERN ISR; (d) dual-solenoid field.

The Split Field Magnet experiment at the CERN ISR, Figure 6.12c, essentially put two horseshoe electromagnets around the interaction point; the proton beam from one direction would bend, then collide with another beam at the IP, then bend back on the way out. The advantages were that particles produced in the forward direction (low p_T particles, diffraction scattering), with a largely p_z momentum, would travel perpendicular to the vertical B_y field and be accurately measured in the tracking system. The weakness was that particles scattered vertically, with p_y, and would travel parallel to the field and be poorly measured.

The dual-solenoid field, Figure 6.12d, has all the advantages (and disadvantages) of the solenoid for tracking, with the added benefit that tracks that emerge from the calorimeter and travel through the annulus between the solenoids have their momenta measured a second time. The weakness of this configuration is that the outer, larger solenoid is necessarily large, although it need not be a precision solenoid. As discussed elsewhere, there are two further advantages to the dual solenoid without iron: the internal forces and distortions of the detector are absent, and the fringe field is essentially zero.

6.4 Tracking System Geometries

The tracking system geometry is determined by the magnetic field geometry so that the momenta of tracks of the most interest are mostly perpendicular to the field, and the sagitta measurement is along the bending plane of the track. In Richter's solenoidal field for Mark I, almost all tracking technologies are easily accommodated. For gaseous wire chambers, the wires are strung axially parallel to $\boldsymbol{B}$ for spark chambers, multiwire proportional chambers (MWPCs), and drift chambers (DCs). A TPC is a natural fit with its strong drift electric field parallel to $\boldsymbol{B}$ ($\boldsymbol{E} \times \boldsymbol{B} = 0$) and with MWPC planes on the end caps. For silicon chambers, the sensor planes and their strips can be oriented in any direction, but the preferred geometry is for the ionization collecting strips to be parallel to $\boldsymbol{B}$. All of these geometries lead to a drift of the ionization electrons perpendicular to $\boldsymbol{B}$, and therefore the specific trajectories of the electron clusters that carry the precision spatial information on tracks undergo the Lorentz force and execute a spiral.

The material in a tracking system is critical for some physics, and the CMS and ATLAS collaborations have both paid careful attention to the distribution of material in radiation lengths (X_0) and, in CMS, also in nuclear interaction lengths (λ_{int}). These are shown in Figure 6.13a–c.

It is clear that the barrel-end cap region near pseudosrapidity $\eta \approx 1.54$ can be a problem for material. The practical need to deliver power and to extract the signals from the inner tracking and vertex chambers take their toll at this point. The implications of these numbers are that track reconstruction codes will deal with electromagnetic showers within the tracking volume from $\gamma \to e^+e^-$ conversions. The rate will be set by the number of π^0s in jets times the conversion probability, approximately X/X_0.

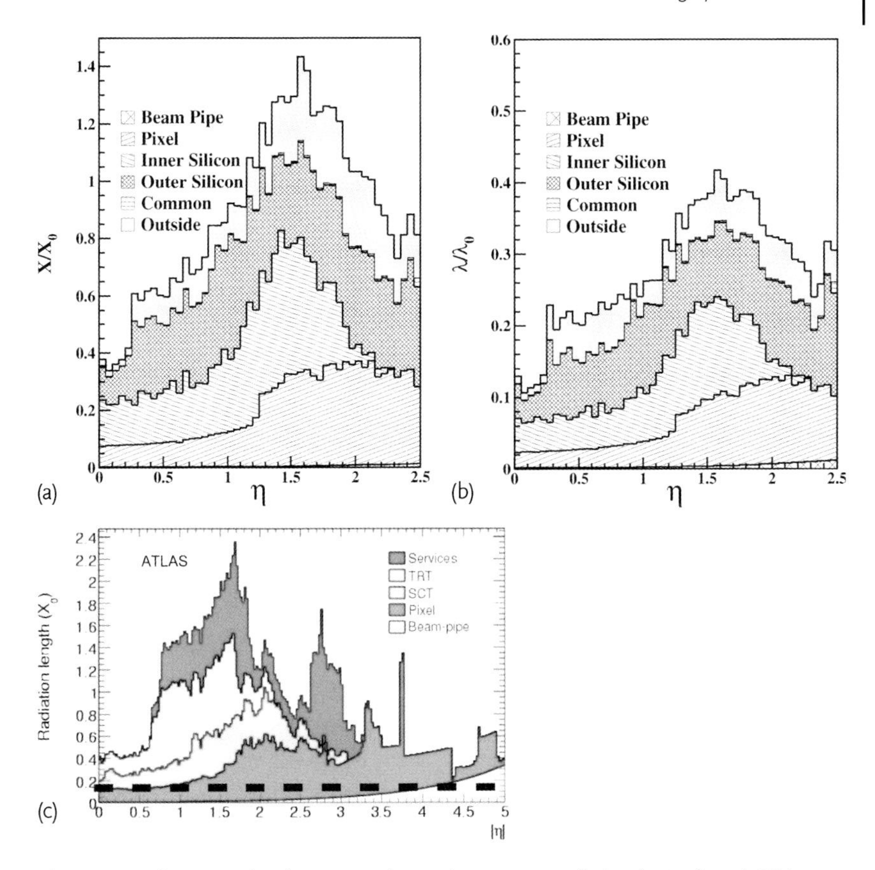

Figure 6.13 The material in the CMS tracking volume in units of (a) radiation length (X_0) and (b) interaction length (λ_{int}) [73]. (c) The material in the ATLAS tracking volume in units of X_0 [74].

For the first time in a big detector, CMS has calculated the number of nuclear interaction lengths in the tracking system amounting to about $0.1\lambda_{\text{int}}$ in the central region, $\eta \sim 0$.

A few years ago there was a proposal for a "spherical detector" (*BILCWS07*, Beijing, 4–7 February 2007) in which the magnetic field was perpendicular to particles from the interaction point everywhere, and so were the silicon tracking wafers. This design maximally took advantage of the cross-product in $\boldsymbol{v} \times \boldsymbol{B}$. The magnetic field design, however, required huge flux densities at the choke points around the beam pipe ($\theta \approx 0, 180°$) on either side of the interaction point in order to obtain a reasonable flux density in the central region at $\theta \approx 90°$. In addition, this field geometry is what a plasma physicist would call a "magnetic bottle", with magnetic mirrors on both ends, for the purpose of *confining* charged particles. Particles in a certain transverse momentum range will execute helices (called loopers) in this field, bouncing back and forth between the mirrors, and eventually lose all their energy through dE/dx in the materials of the tracking system. Although this might optimize the cross-product in $\boldsymbol{F} = e\boldsymbol{v} \times \boldsymbol{B}$, it introduces many technical problems for a practical detector and is not feasible.

6.5 Calorimeter Geometries

Calorimeters have progressed from "shower counters" of stacked Pb-scintillator sandwiched layers to very sophisticated and well-understood instruments. In addition to the physics understanding, the geometry of calorimeters is better understood. There have been many experiments with limited or partial solid angular coverage by calorimeters that result in physics analyses with large corrections for acceptance of final states and, for multiparticle (n) final states in a detector with track or particle acceptance ϵ, the overall event acceptance is proportional to ϵ^n. A lesson for big experiments was learned (and sometimes relearned) that any high quality and high precision experiment must cover the full solid angle with detectors and, in particular, with calorimeters. The obvious exception is the beam hole, plus the necessary access paths for power in and signal cables out. There is as much creativity needed for these problems as for the physics of compensation.

A good calorimeter, either electromagnetic or hadronic, will have an average mass density of 6–8 $\mathrm{t/m^3}$, for example, Pb (11.3 $\mathrm{g/cm^3}$) plus 25% scintillator (1.05 $\mathrm{g/cm^3}$), or U (18.9 $\mathrm{g/cm^3}$) plus 50% Ar (1.4 $\mathrm{g/cm^3}$). A big detector with its calorimeter starting at a radius of 1.5 m will have a calorimeter volume of roughly 100 $\mathrm{m^3}$. Given that all experiments have made the decision to put the massive solenoid outside the calorimeter, the problem of supporting this mass is not easy.

The CMS experiment built the cryostat for its superconducting solenoid with rails on the inside so that the calorimeter could be slid in and supported. The cyrostat contains several centimeters of stainless steel on its inner and outer cylinders, but there is only a small consequence that punch-through particles and muons will lose (an unknown amount of) energy in the materials of the cryostat.

A "projective" calorimeter is now a universal choice, as shown in Figure 6.14.[50] The main point of projective geometry is to concentrate the signal in as few channels as possible to enhance both pattern recognition and to reduce the contribution of electronic noise. The spatial precision of the reconstructed shower is also improved, see Section 3.5.

Forward Calorimeter Region The forward region is a problem for calorimeters as well. At the ILC there is a dedicated group that has designed a series of precision calorimeters [99] that will absorb roughly 3 TeV of electromagnetic *debris* for each beam crossing, while making measurements of the luminosity and the Bhabha rate, and be radiation hard enough to survive years of running.

50) Originally proposed by W. Wenzel, in the unpublished report GIN-59 (Group Internal Note) of 5/23/78.

51) The dose is the maximum conceivable dose due to a full train beam dump onto a Cu plug that appears inside the beam pipe at any position within or near the detector.

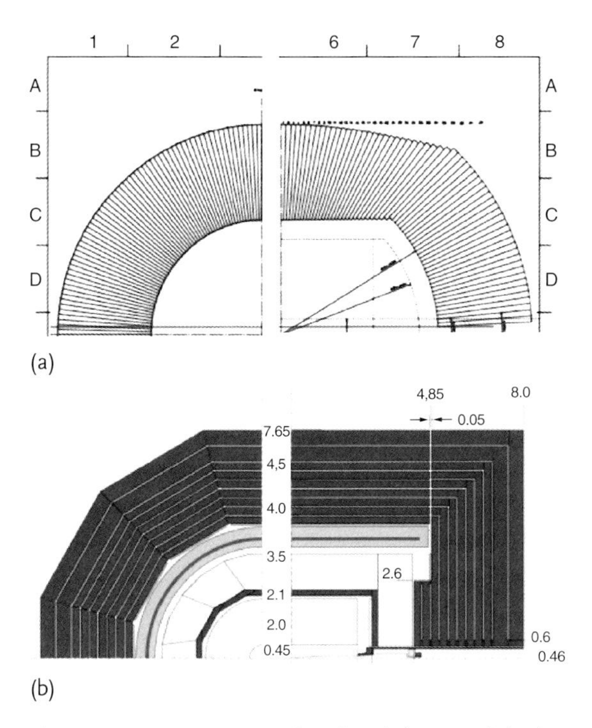

Figure 6.14 Quarter sections of (a) the 4th design and (b) the GLD design of the calorimeter.

6.6 Muon System Geometries

For a solenoidal field and an iron flux return yoke, the natural and easy solution is a complete shroud of iron around the detector, cylindrical in the center region and coming into a pole tip at the openings of the solenoid. The CMS detector is a classic example of this geometry, and also the best designed and equipped muon system in this *genre*. The iron mass serves four good purposes: *(i)* return the magnetic flux, *(ii)* serve as a hadron ($\pi^{\pm}$, K, p, n) particle filter so that predominantly $\mu^{\pm}$ emerge at a large radius, *(iii)* provide excellent mechanical support for all detector systems, especially the massive calorimeter, and *(iv)* serve as a radiation shield, first, for people in and near the IR and, second, for the electronics in the IR. The radiation dose is very large at any proton machine so that even the electronics needs shielding. In an electron machine [19], the dose is lower by orders of magnitude,[51] but in the push-pull schemes for an electron linear collider, in which two detectors will inhabit one IR, the second detector will be on the other side of a shielding wall with personnel present.

A fifth use of an iron muon system is as a crude hadronic calorimeter, as was designed into ALEPH; however, this necessitates that the hadrons penetrate the solenoidal coil. For high precision hadronic calorimetry, the unmeasured fluctuations in energy losses inside the coil are unacceptably large, and no experiment today proposes this.

The identification and reconstruction of $\mu^{\pm}$, with low $\pi^{\pm}$ "punch-through", is limited by the πp and pp elastic and diffractive cross-sections that are about a fraction (0.15–0.25) of the total cross-section. Therefore, an energetic proton can penetrate deep into a calorimeter.

There are two other current muon system geometries: DØ at the Tevatron Collider and ATLAS at the CERN LHC.

DØ Toroidal Iron The DØ experiment has a toroidal iron magnetic field that bends muons axially. The field is limited to the saturation in iron, 1.8 T, and the momentum measurement is made by chambers before and after the iron.

Atlas System of Toroids The magnetic field of ATLAS was described in Section 6.3, but muon identification and reconstruction must go one step further by knowing the line integral of the bending field along the muon trajectory. This field can vary a lot, as seen in Figure 6.3, as muons move out radially.

An interesting comparison will be the quality and character of the muon physics that will be done by CMS and ATLAS in the years ahead.

Iron Ball In the flood of ideas after November 1974, the Iron Ball experiment [104] was designed and built to run at SPEAR with approximately a meter of iron surrounding the interaction point, with provision for small tracking chambers inside in a barrel configuration, and the iron was energized to provide an azimuthal field, B_ϕ, inside the iron. This is akin to DØ, but an extreme case of an experiment that was all muon system and little else.

6.7
Problems and Strategies

6.7.1
Problems with Solenoidal Iron-based Muon Systems

In spite of the preponderance of solenoidal magnetic fields and cylindrical iron-based muon systems, there are problems with this favorite geometry.

$\boldsymbol{B}$ will configure itself so that the total energy stored in the field

$$U = \frac{1}{2\mu} \int B^2 dV$$

is a minimum, and since $\boldsymbol{B}$ is smooth over the volume V, the only big factors are the permeability of the media: for air and other nonpermeable volumes, $\mu = \mu_0 = 4\pi \times 10^{-7} \approx 10^{-6}\,\mathrm{T} \cdot \mathrm{m/A}$, whereas for Fe volumes the stored energy is approximately 1000 times smaller since μ is $1000\times$ larger, $\mu \approx 10^{-3}\,\mathrm{T} \cdot \mathrm{m/A}$.

This drives everything, including all the (large) forces on permeable volumes inside magnetic fields, see Section 6.3.

It is instructive to examine the CMS field in Figure 6.15. A field line, "a Weber", will stay inside Fe for as long as possible, then jump across an air gap to another area of Fe. This minimizes the stored energy of the system

$$U = \frac{1}{2\mu_0} \int\limits_{air} B^2 dV + \frac{1}{2\mu} \int\limits_{iron} B^2 dV \,.$$

When a field line reaches the end of a piece of Fe, it escapes to some distant Fe, such as the elevator, or Fe in the concrete of the IR walls. It jumps from Fe to Fe.[52)]

In Figure 6.15, it is clear that the flux density of 4 T at the end of the solenoid ($z \approx 6$ m) does not all enter the Fe pole tip, which saturates at 1.8 T. Therefore, 2.2 T of the flux density has no energy advantage in entering the Fe pole tip, and prefers to bypass the pole tip and enter the iron support behind it, which with its larger radius has more area. Flux also penetrates the superconducting coil trying to reach the first layer of iron of the muon system, and this flux leads to large compressive forces ($\boldsymbol{F} = I \boldsymbol{L} \times \boldsymbol{B}$) on the solenoid. The flux continues to jump from Fe volume to Fe volume with a small $B^2/2\mu$ stored energy density contribution, avoiding the nonpermeable gaps filled with chambers where $B^2/2\mu_0$ is a thousand times larger. The field is asymmetric top to bottom due to iron members in the concrete floor. The hadronic forward (HF, $z \approx$ 11–13 m) calorimeters are made of iron and draw

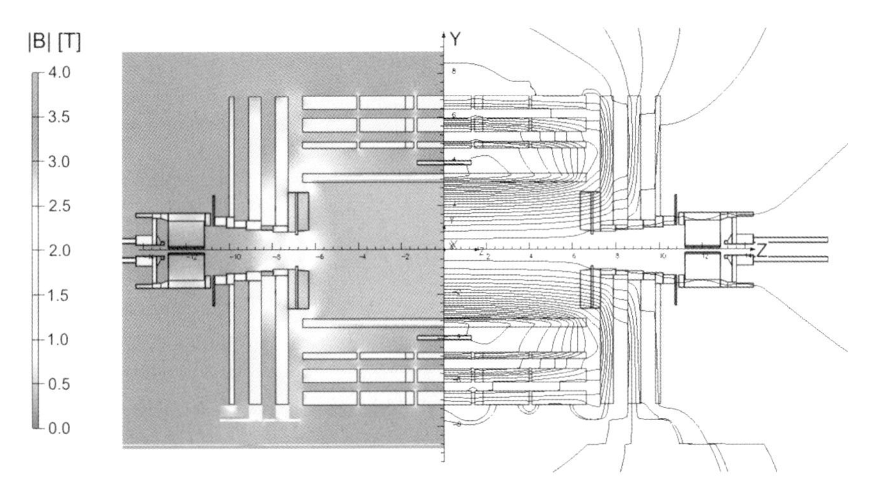

Figure 6.15 The measured magnetic field in the CMS detector, as field intensity on the left and as field lines on the right. This spectacular and informative plot is published in the report "Precision Mapping of the Magnetic Field in the CMS Barrel Yoke using Cosmic Rays", 4 January 2010; authorship is the CMS Collaboration, available at arXiv:0910.5530v2 [physics.ins-det].

52) This provides an easy way to calculate the forces on an iron pole tip such as in this figure. $dU(dz) \rightarrow F_z = -\partial U(z)/\partial z$. dU is simple – $B^2 \cdot dz$.

the flux out axially. The flux at the extremities of the detector, and inside the Fe volumes where presumably the Fe is saturated, leaves the Fe and goes off to find Fe somewhere in the IR. Scientifically, the main point is that the flux density inside the solenoid is highly uniform and excellent for tracking inside this volume packed with precision silicon chambers.

The HF calorimeters moved about 19 mm when the solenoid was energized. This was expected, and it is no problem since one knows where the calorimeters are located to high enough precision. However, for a precision detector at a lepton

Table 6.1 Current and recent detectors and their magnetic fields. Adapted from [49, Table 28.10].

Detector	Laboratory	B [T]	Radius R [m]	Length L [m]	Energy U [MJ]	X/X_0	U/M [kJ/kg]
CMS-2	BINP	1.0	0.35	0.9	0.14	0.376	3.5
CMD-3	BINP	1.38	0.35	0.9	0.26	0.1304	7.0
KEDR	BINP	1.83	1.62	2.84	39	2.483	7.8
TOPAZ	KEK	1.2	1.45	5.4	20	0.70	4.3
CDF	Tsukuba/Fermi	1.5	1.5	5.07	30	0.84	5.4
VENUS	KEK	0.75	1.75	5.64	12	0.52	2.8
AMY	KEK	30	1.29	3.0	40		
CLEO-II	Cornell	1.5	1.55	3.8	25	2.5	3.7
ALEPH	Saclay/CERN	1.5	2.75	7.0	130	2.0	5.5
DELPHI	RAL/CERN	1.2	2.8	7.4	109	1.7	4.2
ZEUS	INFN/DESY	1.8	1.5	2.85	11	0.9	5.5
H1	RAL/DESY	1.2	2.8	5.75	120	1.8	4.8
BABAR	INFN/SLAC	1.5	1.5	3.46	27		3.6
D0	Fermilab	2.0	0.6	2.73	5.6	0.9	3.7
BELLE	KEK	1.5	1.8	4.0	42		5.3
BES-III	IHEP	1.0	1.475	3.5	9.5		2.6
ATLAS							
solenoid	CERN	2.0	1.25	5.3	38	0.66	7.0
barrel toroid	CERN	1.0	4.7–9.7	26.0	1080	(toroid)	
end toroid	CERN	1.0	.83–5.3	5.0	2×250	(toroid)	
CMS	CERN	4.0	6.0	12.5	2600	–	12.0
ILD	KEK/DESY	3.5	3.45	7.3	2000	–	12.2
SiD	SLAC	5.0	2.6	6.0	1560	–	12.0
4th (dual-sol)					2780	0.1	12.0
inner sol		3.5	3.0	11.0		–	
outer sol		−1.5	6.0	12.4		–	

collider, especially one that might need to be periodically repositioned,[53] one does not need any further spatial movements of parts of a detector.

6.7.2
Problems with the Distribution of Material

Material distributed within a tracking system degrades the momentum resolution by introducing a multiple scattering fluctuation into the sagitta (Section 3.2.1) that contributes a term to the inverse momentum resolution such as

$$\frac{\sigma_p}{p^2} \approx \frac{0.016\,\mathrm{GeV/c}}{\ell p}\sqrt{\frac{\ell}{X_0}}[1 + 0.038\ln(\ell/X_0)]\,.$$

One could argue that, beyond measured points at the end of the track trajectory, the amount of material no longer affects the momentum resolution, and this is true, but material begins to hurt the calorimeter performance through fluctuations in the energy losses in this material before the calorimeter. This is more acute for electromagnetic energies. The problem is very similar to the degradation in energy resolution due to leakage in which the leakage is 100%, that is, the rms fluctuation in the shower energy is about equal to the average leakage. For energy lost before the calorimeter, the same is approximately true: the fluctuations in the lost energy are 100%, and so the rms fluctuations in shower energy are about equal to the average energy lost in material before the calorimeter.

6.8
Problems

1. Make a simple cost model for a detector. Start with just a tracking system and a calorimeter, each with reasonable cost per cubic meter. Other than looking up a design report, you can guess a calorimeter is about \$1 M/m^3 and a tracking system is two or three times larger. Your design parameters might track radius and axial extent and the solenoid radius (which limits the sum of the tracking and calorimeter radii).
2. Extend Problem 1 to a detector with a vertex chamber and an iron-based muon system. The costs of pixel detectors are not known, but a good guess is that the cost will be the same for the silicon area, but the electronics costs (which might be 10%) will scale with the number of pixels. You might google the price of machined iron. Also, as a rule of thumb, the human capital costs on a large project are about equal to the capital equipment costs.

53) The ILC, and maybe CLIC, is considering a "push-pull" option with two detectors in one interaction region.

3. Estimate the forces on the pole tips of an iron flux return and, using data on Young's modulus of iron, estimate the spatial deflection of the iron mass. Does this affect the magnetic field in the tracking region?

4. Are there any other possibly useful magnetic field configurations that have not been discussed here?

5. Design a simple toroid for the forward tracking region of a detector that is inside the calorimeter and just outside the main tracker. Take the inner and outer radii to be $r_i = 5\,\text{cm}$, $r_o = 50\,\text{cm}$, and start the toroid at $z = 2\,\text{m}$. Allow yourself 1 m of track path length in a tracking system that is inside the toroid with a sagitta precision of $\sigma_s \approx 200\,\mu\text{m}$. The maximum field at r_i is 1 T. Plot σ_p/p vs. θ, the polar angle of a track. Compare to the θ dependence of the momentum resolution for a conventional forward tracking system, Eq. (3.8).

6. Sketch a redesign of the ATLAS toroids such that the barrel and end-cap toroids are combined into one longer toroid. Is there any way to avoid the highly nonuniform B_ϕ field shown in Figure 6.3?

7. An approximate rule of thumb in Chapter 3 was that calorimeter energies that leak out the side or the back of a calorimeter have a fluctuation that is about equal to the mean leakage, that is, the leakage fluctuates by 100% of itself. If a calorimeter is interrupted in the middle, by mechanical supports, by a solenoid, by cable/power leads, and so on, one might suspect that this "leakage" of energy into a dead volume also fluctuates by 100% of itself. Estimate the constant term incurred by a dead volume of depth $1\lambda_{\text{int}}$ at the center of the calorimeter shown in Figure 3.16 just by looking at these four events. You could, of course, calculate this easily with the GEANT toolkit.

7
4th Concept Detector

"One of the major ingredients for professional success in science is luck. Without this, forget it."

– Leon Lederman

Whenever a new machine is on the horizon, new detectors are designed for it, and in 2005 three detectors were designed [105] for the anticipated International Linear Collider (ILC) [1].[54] In my opinion these excellent detectors, taken together, were too similar to each other: all three were too dependent upon the "particle flow" paradigm, all three had similar iron return yokes, and two of the three had Time Projection Chambers (TPCs) for tracking. I also disagreed with some of the proposed technical solutions, for example, the power pulsing of silicon chamber electronics at 5 Hz to reduce their heat load. I will not assess or evaluate these designs since excellent international review committees [45] have already done so for tracking (Beijing, February 2007), for calorimetry (DESY, June 2007), and for vertexing (Fermilab, October 2007). A 4th detector was briefly proposed [106] as an option and as an alternative especially in the area of hadronic calorimetry. This was a low risk proposal since a thorough beam test of a prototype dual-readout calorimeter had already been completed at CERN and the results published [52, 107, 108], and, in addition, Wigmans had given a talk [105] at this same meeting on compensating calorimetry, *viz.*, the achievement of the ILC calorimeter goal of $30\%/\sqrt{E}$ energy resolution in the compensating SPACAL calorimeter.

I will outline the design of this "4th" concept[55] detector in many particulars: tracking and calorimetry, magnetic field and muon detection, machine-detector interface and beam-beam stability, physics reach, and scientific flexibility. It is a design that is not finished and certainly not engineered. For example, we do not have an elegant solution to the problem of tracking in the very forward direction, nor do we have a pixel vertex chamber technology, and we have not explored the inter-

54) There is a long history of this machine, starting with the first prototype linear collider, the Single Pass Collider (SPC), or SLAC Linear Collider (SLC) at SLAC, followed by the Japan Linear Collider (JLC), the Next Linear Collider (NLC), and the TeV Energy Superconducting Linear Accelerator (TESLA).

55) "Concept" in this context means this detector that does not exist but is in the conceptual design stages.

Particle Physics Experiments at High Energy Colliders. John Hauptman

ISBN: 978-3-527-40825-2

play between vertex chamber and the main tracking chamber. The 4th is different in almost every respect from the other three ILC concept detectors, but it is not deliberately different.

7.1 Description of the 4th Detector

The 4th detector is designed to meet the challenge of a high precision detector at a high energy e^+e^- collider in the $\sqrt{s} \sim 0.5$–$1\,\text{TeV}$ range with energy, momentum, and spatial resolutions that are two to ten times better than the four already excellent LEP detectors. We believe that fundamentally new ideas are required to get there, in particular, (a) a new low-mass tracker, (b) dual-readout calorimetry, and (c) a dual solenoid for iron-free flux return. The 4th concept consists of these three innovations plus a vertex chamber, and besides these there are several more innovations mostly in particle identification techniques.

It is not clear that this design, either in its particulars or in its generalities, is adequate to the task of a precision detector at a future collider. In my opinion, it is not clear that any presently proposed detector can meet the requirements needed to achieve all the benchmark physics processes that must, at a minimum, be demonstrated. The physics problems are easy, at least compared to the background problems. Of course, all proponents say otherwise, but short of extensive beam tests of all components and subsystems, including some accounting of track backgrounds that we already know will exist, we may not have the certainty and finality that a project of this magnitude requires.

1. A pixel vertex detector is required for high precision vertex definition, impact parameter tagging, and near-beam occupancy reduction. The specific impact parameter resolution that must be achieved is

   $$\sigma_b \approx 5\,\mu\text{m} \oplus 10\,\mu\text{m}/p \sin^{3/2}\theta\ .$$

   Like all tracking systems, the success or failure of a vertex chamber will depend upon the backgrounds close to the beam [45, Vertex Review]. The problem of a $(20\,\mu\text{m})^2$ pixel chamber 1–2 cm from an electron beam, with low mass (less than $0.1\,X_0$) overall and with readout on the microsecond scale, is probably the hardest problem facing any new detector. Many talented groups with many ideas are working on pixel vertex chambers [47].
   The relationship between the pixel vertex chamber and the main tracking chamber is conventional, that is, we have not explored designs in which the vertex chamber is reduced in scope to a single-function impact parameter tagger, nor have we considered a much larger chamber.
2. A cluster-timing and cluster-counting ultra-low-mass drift chamber with a He-based gas mixture and about $N = 150$ three-dimensional space points along a track. The spatial resolution perpendicular to the wire is estimated from tests and calculations to be about 55 μm per point. This "continuous" tracker has the

advantages of a continuous-tracking TPC with most of its extraordinary pattern recognition capabilities and, with its short drift distance of $r \approx 0.5$ cm, the advantages of a silicon tracker with a readout time within one beam crossing (337 ns at an ILC). The large number of independent point measurements makes the chamber less reliant on both the efficiency of single points and the alignment precision of a small number of points at the 5-μm level. The required raw spatial resolution of 5 μm is recovered as 55 μm/$\sqrt{N}$. In addition, a counting of electron ionization clusters, which are Poisson distributed without a Landau δ-ray high-side tail, opens up the possibility of a 3.5% specific ionization measurement for particle identification in the few-GeV region.

The weakness of a drift chamber is that the volume of one channel is large, $\pi r^2 L$, so that in a high track background environment the channels at low radius can be filled at every crossing (high "occupancy") and the channel effectively disabled. This chamber, called "CluCou", avoids the problem of positive ions and the single-bunch crossing readout by keeping the drift distance small, about 5 mm. This chamber is not as good as a TPC for pattern recognition and not as good as a silicon system for timing and single-point precision, but it is nevertheless adequately good for momentum resolution and single-crossing readout.

3. A high precision dual-readout fiber calorimeter, complemented with an EM dual-readout crystal calorimeter, both with a time-history readout, for the energy measurement of hadrons, jets, electrons, photons, missing momentum, and the tagging of $\mu^{\pm}$ and $\tau^{\pm}$ (including $\tau \to \rho\nu$). There is very little left for speculation on dual-readout calorimeters since so much data have been published.[56)] On the other hand, the path to a big detector based on dual-readout calorimetry is not clear, either. Although fully and honestly simulated in the 4th letter of intent [75], the problem of a photoconverter was not solved, nor was that of the cableways for taking out the signals.

   It was our decision to put dual-readout crystals in front of the dual-readout fiber calorimeter, primarily to ensure that detailed and high precision electromagnetic measurements, Section 4.1.2, would be achieved, and specifically to assess this we used the important $\tau^{\pm} \to \rho^{\pm}\nu$ decay as a measure, Section 4.5. However, it is not clear that crystals are a big help. They require clever mechanical support[57)] within the shell of the fiber calorimeter and cableways to get the signals out, which results in dead volumes in the fiber calorimeter. One option is to increase the Čerenkov photon yield in the fibers so that good electromagnetic energy resolution ($10\%/\sqrt{E}$) is achieved in the fiber calorimeter alone, and to make the lateral segmentation of the fiber channels smaller, almost as small as the crystals. A cost-benefit analysis might even find that a calorimeter system without crystals is less expensive, and a physics assessment might show that most of the physics capabilities in particle identification, calibration, and

56) A selection of the 18 DREAM papers that are relevant to a big detector are those on the measurements of hadrons [52], electrons [108], muons [107], neutrons [120], lead tungstate as a calorimeter [53], and BGO crystals as a dual-readout calorimeter medium [54].

57) We adopted the solution in the BaBar experiment by A. Miccoli, INFN, Lecce.

background rejection are maintained. In other words, a full exploration of all calorimeter configurations has not been done.

4. A dual solenoid to return the flux without iron and provide a second field region for the inverse direction bending of muons in a gaseous tracking volume to achieve high acceptance and good muon momentum resolution and identification. There are many secondary advantages to an iron-free detector such as noninterference with the magnetic elements of the final focus, machine-detector interface (MDI) (Section 7.2) advantages, a very lightweight detector in the IR, and zero fringe field. From the accelerator point of view, these might be called primary advantages.

The pixel vertex chamber, tracking, and calorimeter are inside the solenoidal magnetic field with a radius of about 3 m.

Each of these detector systems is self-contained and makes its measurements without assistance from, or dependence on, any other system of the detector. This is the first principle in Section 6.2.

In addition to the primary measurements, each system makes independent measurements that lead directly to particle identification, and in all cases it is a combination of two or more measurements that provides good particle identification, sometimes in the same detector system (specific ionization dE/dx and momentum p) or in different systems (energy E and momentum p for $e^{\pm}$ identification).

As examples of particle identification, we achieve a reconstructed jet energy Gaussian resolution in the dual-readout calorimeters of $\sigma/E \approx 29\%/\sqrt{E} \oplus 1.2\%$ on complete event ensembles without prior selections of any kind. This leads to $W \to jj$ separation from $Z \to jj$ by mass, and high precison missing ν three-vector by subtraction. The dual-readout fiber calorimeter measurements lead to $e - \mu - \pi$ separation and tagging, n ("neutronic" or "hadronic") tagging, and further EM-hadronic separation by the time structure of the signals. The dual-readout crystal calorimeter yields high precision EM measurements for $e - \gamma$ separation and $\pi^0 \to \gamma\gamma$ reconstruction, and the prompt Čerenkov light yields precision time measurements for time-of-flight separations. In addition, the cluster-timing chamber achieves a 3.5% specific ionization measurement for $e - \mu - \pi - K - p$ identification in the few-GeV region. All of these capabilities are integrated into this one detector.

The 4th detector was initially designed around a dual-readout fiber calorimeter as previously tested by the DREAM collaboration [52]. The subsequent evolution of the detector and its components was guided by the principles listed in Section 6.2, which we list here again annotated for the 4th detector.

1. *Each detector subsystem is independent*, and the essential measurements of the calorimeter, for example, do not depend on the tracking system or any other system.

   The 4th calorimeter is explicitly independent of any other system, and a complete physics program could be envisioned without a tracking system at all.

The tracking cluster-counting chamber will find tracks and the decays of tracks and neutrals within the tracking volume without assistance from the vertex chamber or calorimeters. The same is true of the tracking annulus between the solenoids that measures the momenta of all tracks that pass through the calorimeter and solenoid (or are generated by hadronic interactions in their materials). If there is any *reason* for this independence principle, it is that no detector system is perfect, especially in large detectors, and one does not want one poorly performing detector to compromise another one that might have been a good detector. Another reason might be absolute calibration of energies or momenta in which correlations can confuse the issue.

2. *Particle identification is critical to all physics,* and careful attention was paid from the beginning of the design to not only achieve particle identification measurements, but to see that one detector system did not compromise another.

   If we knew the identities of every particle in an experiment, physics analyses would be almost trivial. The critical quantities are the efficiency and the corresponding purity of an identified sample of objects. As one makes a selection in some identity parameter, for example, the impact parameter significance in Figure 3.12b at a value of $\log(1 + b/\sigma_b) = 4$, the b-jets outnumber the c-jets, but not all b-jets are accepted and not all c-jets are rejected. At a selection of $\log(1 + b/\sigma_b) = 6$, the purity of b-jets is much higher, but the b-jet efficiency is much lower. Such a choice is inevitable in all selections against backgrounds and the situation is only improved by precision measurements.

3. *Auxiliary or ancillary detectors should be unnecessary.* This directly implies that each independent detector system must be very capable, and runs contrary to the practice in many existing experiments.

   This might be a contentious point, but it makes sense to me to put more interaction lengths of material into the hadronic calorimeter, for example, than to build a separate "tail catcher" system behind the coil. Every new detector system, however small or simple, requires its own power facilities, cableways, and supports. If one system can be designed to do all the measurements, even if it is larger and more expensive, this is better than two systems. This runs counter to the political needs of multiple groups each requiring its own system (its own "territory") within a large detector (Section 8.5).

4. *Use common technologies whenever possible,* such as the same chamber technology or the same electronics or power supplies for multiple detector systems.

   This can be seen as an economy measure, but it is more than that. Calibration and monitoring continues throughout the life of a detector and they require human time and attention. Common systems not only for detectors, but also power supplies and electronics, reduce overall labor. Another way to phrase it is, if you have an excellent tracking or calorimeter technology, why not re-use it as much as possible, with modifications for regions of the detector with different backgrounds or particle rates.

5. An *iron-free detector and dual-solenoid flux return* opens up many new opportunities in detector design, while simultaneously avoiding some of the evident problems of iron-based detectors such as overall mass, internal forces, and fringe field.

   An iron-free detector has many heretofore unrecognized advantages that cut across many aspects of a detector, including precision momenta after the calorimeter coil (for physics), low mass in the IR (for civil engineering), no fringe field (for machine optics), and flexibility for purely speculative long-lived weakly decaying states at 5–7 m from the IP. A detailed discussion of these is given in the machine–detector-interface section, Sections 7.1.4–7.2, but the implications of a detector without an iron yoke go far beyond the interferences between detector and machine. The idea of using multiple solenoids to funnel the magnetic flux around in a detector was originally conceived by A. Mikhailichenko [109], and he generated the 4th design shown, for example, in Figures 6.5 and 7.13, in which the funneling is almost perfect with zero resulting fringe field and a useful uniform magnetic field throughout all of the detector volume (except inside the calorimeter where you neither want nor need it).

6. *Keep watching each and every channel in the time domain* for as long as possible, and with time slices of 1 ns or less.

   Fast analog-to-digital converters (ADCs) do not yet exist at costs compatible with a big detector, but many groups are working on these and there should be a "good market" for them. Many of the particle identification measurements and precisions of the 4th depend on what we call "time history". The electronic wire is already there and memory is cheap, so just keep watching and recording the signal. One of the first experiments to clock out the time history of a channel, and therefore the 6-m^3 tracking volume, was the Berkeley TPC by Nygren, which took data at SLAC/PEP4 for many years. The benefits of a whole-volume interrogation of the entire track ionization led to many strengths, including the specific ionization measurements in Figure 3.1a and spatial resolutions in $r\phi \approx 100\,\mu\text{m}$ and $z \approx 500\,\mu\text{m}$ that exceeded the goals set forth in the proposal. A billion dollars' worth of TPCs have since been built worldwide. The drift of ionization electrons in the Ar$-$CH$_4$ gas mixture was slow by the necessities of future colliders, and fast digitizers do not yet exist to accomplish the idea of GHz digitization of all detector channels of a big detector, but it would be directly useful in particle identification (Chapter 4) and its potential for background suppression, both for collisions and in between collisions, has not yet been explored.

7. The *relative absence of inactive, or dead, regions within the instrumented volume.* This is an obvious statement and no one wants a partially inactive detector.

   Dead regions are unavoidable for the support of massive instruments such as the calorimeter and the superconducting coils, but it is quite avoidable in calorimeters in which all readouts can be at the rear, and for lightweight tracking systems that do not require cooling or fragile supports within the tracking volume itself. This problem is difficult and, maybe, outside the control of the dreams of physicists, given Young's moduli of useful materials. However, new materials and composite materials should be watched closely for breakthroughs in density and strength. There is no substitute for the support of a solenoid or a calorimeter, and on the 4th we have tried to introduce new ideas and materials into the support of the calorimeter and tracking chamber as discussed in the talk by Grancagnolo [110].

8. A final strength of a big detector (especially one built under scheduling or funding duress) is a design that is *resistant to "engineering creep"*.

   I do not know how else to phrase it, but all detector designs, TDRs, and even engineering designs lack the final specific engineering details. It is these unanticipated additions that can compromise performance of a detector. I am not even sure what this means, but it is a real effect. It has to do with the transition from plans and calculations to actual items in a machine shop, to the final testing and assessments of cooling needs and mechanical and stability supports. All of these small augmentations add up, and problems are almost always solved by adding material to the detector. Furthermore, after a technology has been chosen and the collaboration is far down the path of building it, switching to another technology is not possible for both scheduling and funding reasons. This fact only emphasizes the importance of thorough beam tests of all detector components.
   But the tests themselves are not enough. Physicists must honestly assess the results of beam tests, and make the much more difficult extrapolation to a large system embedded within a large detector. They must also be willing to reject a favorite detector when it fails testing.

Therefore, a good detector should allow all particles to pass through a very low mass tracking region in which only the slightest ionization signals are extracted for sagitta measurements, then enter a very high density calorimeter volume that only muons and neutrinos can penetrate, and both tracking and calorimeter volume elements are read out each nanosecond. We have designed the 4th to be as close as reasonably possible to this ideal detector, based on existing detectors and beam tested prototypes whose performance we understand and propose to enhance: for example, the KLOE tracking chamber augmented to cluster-timing CluCou [42], beam tests of the DREAM augmented with time-history readout, or existing superconducting solenoids (KEDR AND CMD-2M solenoids) built with new techniques [101, 111].

Finally, a good detector should be balanced in its physics measurements, that is, given that $W^{\pm}$ and Z^0 decay to all their final states are important, the measurement capabilities of the whole detector should measure each and every one of these final states with comparable precision. This has only become possible with the development of high precision hadronic calorimetry, and this allows us to begin

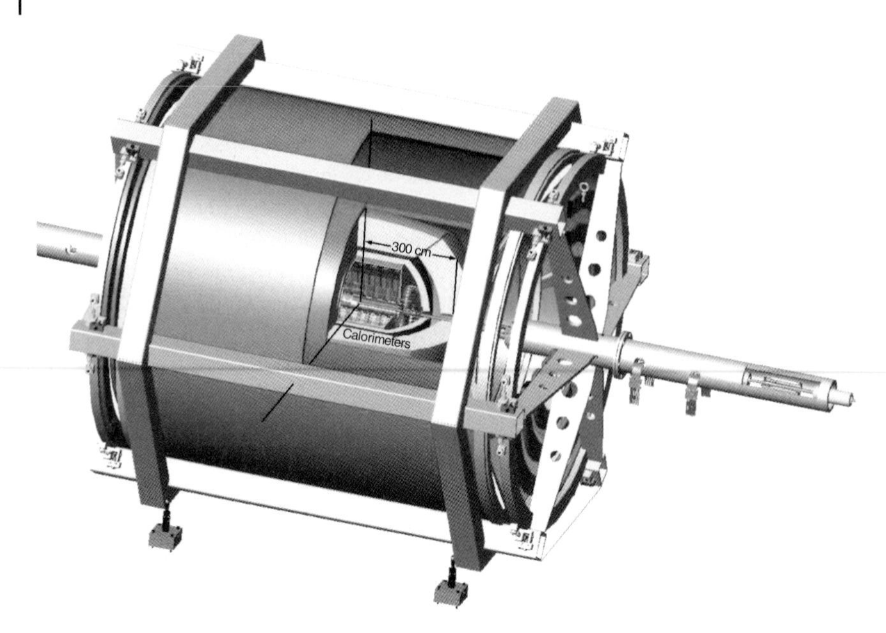

Figure 7.1 The 4th detector showing final focus transport, the dual solenoids for iron-free flux return, the vertex and tracking systems, and the calorimeters: the inner one is a dual-readout crystal and the outer one is a dual-readout fiber, both with time-history readout. The total depth is $10\lambda_I$ and all calorimeter channels are projective with the origin. The frame is nonmagnetic and easily able to contain the magnetic pressure $\propto B^2$. The gross dimensions are 12 m diameter and 16 m length, excluding the beam delivery.

to think of detectors with energy and momentum resolutions for $e^{\pm}, \mu^{\pm}, \gamma$, and j that are comparable. Such a detector would be truly "general purpose".

7.1.1 Gross Design

We first discuss in general terms the radial (r) and axial (z) extents of the detector, and the gross dimensions of the tracking and calorimetry. In Sections 7.1.2 and 7.1.3 we connect the detailed beam test data and R&D to the 4th design, Figure 7.1.

The r-Extent of the Detector The DREAM test module is $10\lambda_{\text{int}}$ in depth, sufficient to suppress depth leakage from single pions and nucleons at the highest energies, and more than sufficient for jets [52]. In a 4π detector we seek to maintain this depth not only for hadron measurements, but also as a further pion filter to assist in muon survival identification and to serve as additional radiation shielding for mdi purposes. These considerations lead to a 1.5-m fiber calorimeter depth.

The momentum resolution requirement that $\sigma_p/p^2 \sim k(\sigma_x/\sqrt{N})/B\ell^2 \sim 3 \times 10^{-5}$ (GeV/c)$^{-1}$ demands a small space-point resolution at many points, $\sigma_x/\sqrt{N}$, and large $\boldsymbol{B}$ and ℓ^2. The field strength $\boldsymbol{B}$ is largely governed by the superconducting current density limit and the stability and stiffness of a large solenoid and will fall within the range $B \approx 3$–$5\,\mathrm{T}$. It is easier to gain momentum resolution from ℓ^2 at the expense of a larger calorimeter volume and a larger solenoid. We can achieve this resolution in CluCou with an outer radius of the tracking system at $R \sim 1.5\,\mathrm{m}$. For a 30-cm deep crystal calorimeter before a 1.5-m-deep fiber calorimeter, the inner solenoid starts at $R_1 \sim 3.3\,\mathrm{m}$. Using CMS numbers [100] for conductors and current densities, it is easy to design a dual solenoid resulting in a uniform $B_z = 3.5\,\mathrm{T}$ in the tracking volume and $B_z \approx -1.5\,\mathrm{T}$ in the annulus between the solenoids with an outer-solenoid radius of $R_2 \sim 6\,\mathrm{m}$. This radius is not critical for either current densities or muon momentum resolution. The currents in the outer solenoid are modest compared to those of the inner solenoid, and therefore the outer solenoid is "easier" than the inner solenoid, except for its size.

The momentum resolution of muons in the annulus is approximately independent of the outer solenoid radius since the flux density through the inner solenoid is $B_1 \approx 3.5\,\mathrm{Weber/m^2}$, or $3.5 \times \pi R_1^2$ Weber, and these Webers fill the annulus area $(\pi R_2^2 - \pi R_1^2) \sim \pi R_2^2$, where the flux density is $B_2 \sim 3.5\pi R_1^2/\pi R_2^2 \sim 1.5\,\mathrm{Weber/m^2} = 1.5\,\mathrm{T}$. If R_2 is made larger, the field decreases as $1/R_2^2$ while the track length squared $(\ell = R_2 - R_1)^2$ increases as $(R_2 - R_1)^2 \sim R_2^2$. These factors cancel, $B_2\ell^2 \sim$ constant, and the momentum resolution is independent of R_2. Therefore, we choose R_2 to be reasonable, say, $R_2 \sim 6.0\,\mathrm{m}$. If we made R_2 substantially smaller, we would incur larger current densities, a smaller B_1, and more difficult construction tolerances.

In fact, making R_2 larger lowers both the currents and the field for the same flux and makes it possible to increase the central tracking field to 4 T. We have worked on designs for large solenoids [111] but have not proceeded to engineering work.

The z-Extent of the Detector The extension of this detector in z is not a serious design problem for the end coils of the dual solenoid, the axial muon spectrometer, the end calorimeter, or the end caps of the tracker, except for the wire length in CluCou. None of these are high cost, and one does not affect the other. The cost increases are predominantly associated with a calorimeter of a larger volume, whereas the chambers are small cost by volume. Therefore, we choose the end-cap calorimeter to begin at $z \approx 2.0\,\mathrm{m}$, so as to not complicate the construction of the tracking chamber, and the magnetic volume to end at $z \approx 7.5\,\mathrm{m}$.

Gross Calorimeter Geometry The DREAM fiber module [52, 107, 108] is a good model we choose to follow with a $10\lambda_{\mathrm{int}}$ depth and a 22% fiber volume fraction. The crystal BGO calorimeter tested by the DREAM collaboration is also a good model since the data are unambiguous [54]. The depth of the BGO must be at least $25\,X_0$, and this we choose as sufficient depth to suppress electromagnetic leakage fluctuations while not accepting too large an optical attenuation in the crystal, which would lead to a constant term in the electromagnetic energy resolution.

Gross Central and Muon Tracking The data from the KLOE chamber demonstrate a chamber with excellent long-term mechanical and electrical stability. There are essentially no restrictions on radius or length, except for gravitational sagging and electrostatic wire instabilities for longer wires. We studied the use of low-mass wires that can be strung at sufficiently high tension to alleviate both of these concerns. Therefore, the tracking chamber geometry is not a design concern.

The same wires and cluster-timing electronics used in the main tracker are used in the muon spectrometer in the annulus between the solenoids, except that the wires are inside precision tubes as in the ATLAS experiment. The volume is much larger, but the length of the tubes is kept below 4 m.

Therefore, the gross design of the detector is set. The simulations of events are done in this geometry. We have not done what could be called a "detailed optimization" of the detector.

7.1.2
Tracking Systems: Pixel Vertex and Main Tracking

A tracking system really means everything between the interaction point and the calorimeter: the beam pipe, the vertex chamber, and the main tracking chamber. The beam pipe is a straight beryllium cylinder with an inner radius of 1.2 cm and a wall thickness of 0.04 cm.[58)]

The Vertex Detector is a multi-gigapixel chamber with five cylindrical layers and four axial disk layers. Several groups are designing such detectors, and in our simulation we use a generic[59)] "thin pixel" design. It is our expectation that this chamber will achieve an impact parameter resolution of

$$\sigma_b \approx 5\ \mu\text{m} \oplus 10\ \mu\text{m}/p \sin^{3/2}\theta \oplus 10\ \mu\text{m}/\sqrt{p}$$

with square pixels approx. 20 µm in size.

The total budget of the vertex chamber and beam pipe is 1.2% of X_0. The performance of this vertex detector in ILCroot with and without beam background is described in [75].

The tracking chamber is an ultra-low-mass drift chamber modeled on the successful KLOE chamber at Frascati, already the largest and lowest mass chamber ever built, and filled with a He-based gas to reduce by a factor of 10 both the multiple scattering of physics tracks and the electromagnetic conversions of beam-crossing debris in the tracking volume. We have made two augmentations [42] to this chamber:

1. clock out each wire at $\sim$ 1 GHz and count individual clusters of electrons, providing a 3.5% measurement of specific ionization (without a Landau tail) for particle identification in the few-GeV region, improved spatial resolution by in-

58) A conical profile to avoid the envelope of background tracks leads to a detailed design of the beam pipe given in the thesis "Two-Jet Analysis of Physics Events at the International Linear Collider" by A. Mazzacane.

59) Similar to the pixel vertex chamber of the concept detector SiD.

cluding all clusters in the determination of the impact parameter of the track, a dip angle measurement and therefore a crude z measurement of each cluster, and a total drift time less than the bunch crossing time of 337 ns; and
2. new composite wires (made in industry) of a high tensile strength 20-μm-diameter core, coated with Ni and overcoated with Ag for conductivity. These small wires reduce the multiple scattering so that it is comparable to the extremely low multiple scattering in the gas itself (the irreducible multiple scattering), allowing lower operating voltages, and therefore easier electrostatic instabilities, and lower bulkhead forces and mass. The momentum resolution of the tracking chamber is

$$\frac{\sigma_p}{p} = \frac{\sqrt{320} \cdot \sigma_{r\phi}}{0.3 \cdot B \cdot \ell^2 \cdot \sqrt{N}} \cdot p \oplus \frac{5.4 \times 10^{-2}}{B \cdot \ell} \sqrt{\frac{\ell}{X_0}} \,,$$

with $\sigma_{r\phi}, \ell$, and X_0 in [m], B in [T], and momentum in [GeV/c].

We were not confident about a 5-layer, high precision silicon strip tracking chamber for several reasons: (i) there are too few points on a track, so that background hit confusion, inefficiencies, and overlapping tracks could lose a hit, reducing a track to only four points in the main tracker (plus those in the vertex chamber); (ii) to keep the multiple scattering mass down, liquid cooling must be replaced by gaseous cooling complemented with power-pulsing[60] at 5 Hz, with its periodic Lorentz forces, $\boldsymbol{F} = i\boldsymbol{L} \times \boldsymbol{B}$, on all conductors carrying current i that can lead to the enabling of mechanical harmonics of this precise 5-Hz driving frequency and endangering the delicate bump bonds throughout the chamber; and (iii) calibration to 5 μm is difficult.

We were not confident about using a TPC, in spite of its beauty and power at tracking anything in almost any environment, because (i) the density of the gas (usually Ar) results in multiple scattering, (ii) the buildup of positive ions in the gas volume leads to a radial electric field component and alter the $\boldsymbol{E} \times \boldsymbol{B} = 0$ requirement of high precision tracking, and (iii) the end walls with their sophisticated electronics with a moderate fraction a radiation length of material.

The cluster-counting helium drift chamber has a factor of 7 less multiple scattering and a slower drift velocity, making the counting and the timing of the individual ionization clusters easier, but still collecting all the ionization clusters within one beam-crossing time. In addition, helium has a much smaller photoionization cross-section for the photons coming from synchrotron radiation and beamstrahlung and is relatively transparent to this debris field.

Cluster Timing and the Ultimate Resolution Drift Chamber In a He-based gas, the individual ionization clusters impinging on the sense wire are separated in time by a few nanoseconds to a few tens of nanoseconds. These long times are due

60) The particular beam train structure of ILC has five approximately 200-ms periods per second without beam bunch crossing, and one turns off the power to the tracking system during these periods.

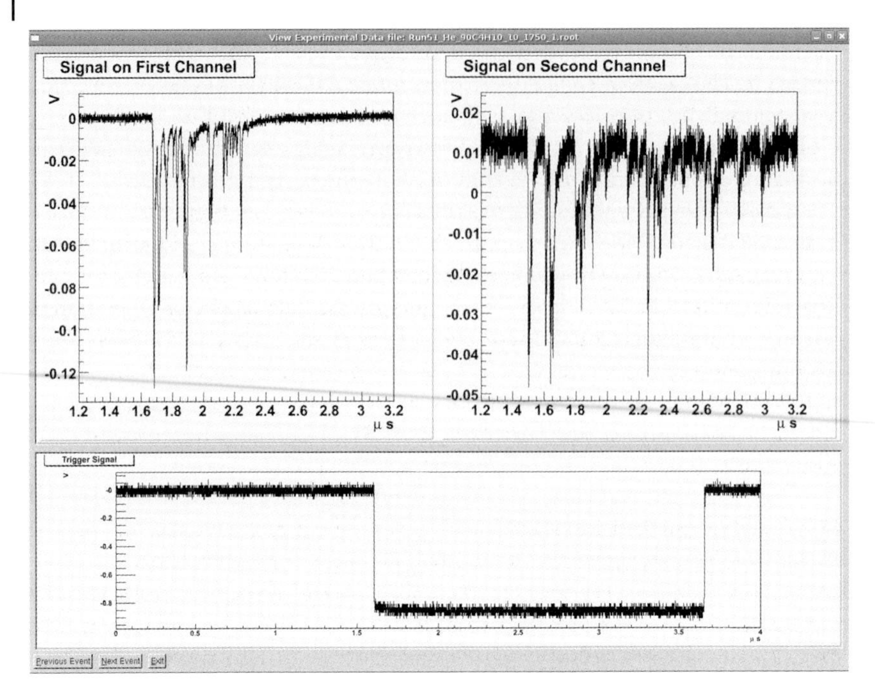

Figure 7.2 Digitized pulse shape showing individual clusters in two separate test chambers with different cell sizes. Gas is 90% helium–10% isobutane at NTP.

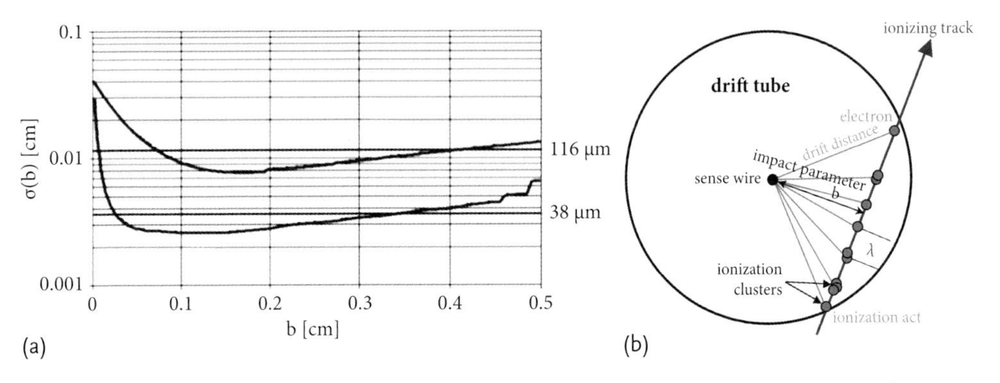

Figure 7.3 (a) Spatial resolution as a function of the impact parameter for a 0.5-cm cell and a helium-based gas mixture. The upper curve represents, for smaller *b*, the overestimate of the impact parameter when it is assumed to be equal to the distance of the first arriving electron. The lower curve shows the impact parameter resolution calculated by taking into account the arrival time of all ionization clusters. The straight lines are the average values of the two described resolution functions over all impact parameters. (b) Schematic of individual clusters on one wire.

to the low density of ionizations per centimeter in He (a factor of 5–8 less than Argon) and the slower drift velocity in He (a factor of 2–3 less than Argon). At these times, efficient electron cluster counting can be performed with a GHz, or slightly faster, digitizer. Clusters in a test chamber triggered on cosmic rays are shown in Figure 7.2. The expected density of primary ionization clusters in this gas mixture for a mip is 12.5 /cm, with a total number of electrons of about 20 /cm. The time of the first cluster, t_1, is almost right, but imprecise by the fluctuations relative to the shortest perpendicular path to the wire, as seen in Figure 7.3b.

The main fluctuations affecting the impact parameter coordinate resolution are (a) the ionization statistics in the gas, (b) electron longitudinal diffusion on the way to the sense wire, (c) the time resolution of the electronics, and (d) the time-to-distance relationship. The largest contributors are the fluctuations in the ionization statistics, and, by counting the separate clusters as they arrive at the sense wire, this fluctuation can be reduced. In fact, the second cluster provides about half the improvement, and the next several clusters the remaining improvement. In the end, the spatial resolution on a track, Figure 7.3a, is about 116 μm using the first cluster alone, and improves to 38 μm using the first several clusters.

dE/dx, dN/dx and Particle Identification Counting ionization clusters has further benefits. Measurements of ionization energy loss, Figure 3.1, have large "Landau fluctuations" in the energy deposited in a gas by a charged particle due to the production of δ-rays or energetic electrons. In order to calculate the mean ionization loss, a traditional truncated mean algorithm is used that throws out the 20% or 40% of the highest ionizing samples, and the mean of the remainder has a much smaller fluctuation. Typically, a dE/dx resolution of 6% is achieved, as in Figure 3.1, where 4.5% is the theoretical optimum. Counting clusters, however, is Poisson and with a sample of $N \sim 10^3$ clusters (12.3 cm × 130 cm), the dN/dx resolution can be $1/\sqrt{N} \sim 3\%$.

7.1.3
Calorimeters: Fiber and Crystal Dual Readout

The calorimeters of the 4th design are tested and proven technologies, fully published,[61] and widely detailed in numerous talks at international meetings. Parallel to the developments in instrumentation, we have developed a full and detailed simulation of dual-readout calorimeters in both fibers and crystals [112] in ILCroot that is faithful to the DREAM beam test data.

It was not obvious that a combination of two calorimeter technologies, although both dual-readout capable, would result in an exceptional (or even adequate) calorimeter system for physics, but this combination works very well. This is a stunning achievement of the DREAM collaboration and Di Benedetto [112] to show that, taken singly, both crystal and fiber dual readout work and then, taken together, that a full-size BGO calorimeter in combination with a $\sim 9\lambda_{\rm int}$ fiber calorimeter also works perfectly well [55].

61) The DREAM collaboration includes N. Akchurin, F. Bedeschi, A. Cardini, R. Carosi, G. Ciapetti, R. Ferrari, S. Franchino, M. Fraternali, G. Gaudio, J. Hauptman, M. Incagli, F. Lacava, L. La Rotonda, S. Lee, M. Livan, E. Meoni, A. Negri, D. Pinci, A. Policicchio, S. Popescu, F. Scuri, A. Sill, G. Susinno, W. Vandelli, T. Venturelli, C. Voena, I. Volobouev, and led by R. Wigmans. The main DREAM papers are [52], [53], [54], [56], [107], [108] and [120]

(a)

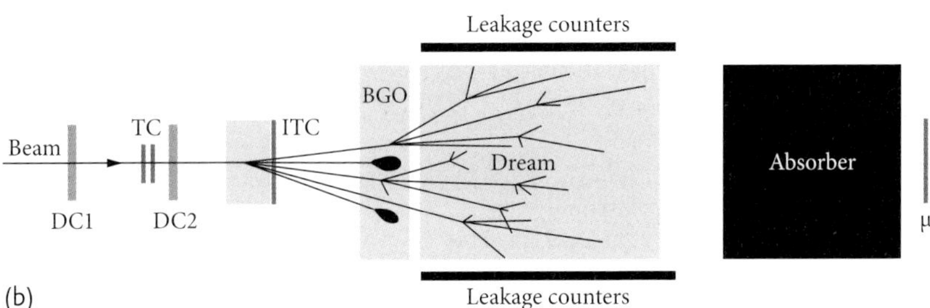

(b)

Figure 7.4 (a) Dual-readout crystal and fiber calorimeter tests at CERN (July–August 2008). The beam ($e^-, \pi^\pm, \mu^\pm$) enters from the left in this image, encountering $6 \times 6\,\text{cm}^2$ trigger counters, an xy-beam chamber, the 100-crystal BGO array shown, followed by the DREAM module, which is surrounded by eight large scintillating paddles to sample the leakage from the DREAM module. Beams from 20 to 300 GeV were used in this test. This configuration is close to the 4th proposed arrangement of a crystal dual readout in front of a deep fiber dual-readout calorimeter (but with the crystals aligned with the beam) and allows us to practice calibrating the fiber calorimeter behind the crystal calorimeter and to intercalibrate the two detectors. This work is done by the DREAM collaboration led by Richard Wigmans. (b) A sketch of the above setup. [BGO-Leak-DREAM].

The physics motivation for this combination is to achieve excellent electromagnetic four-vector resolutions on γ and $e^\pm$ from decays of any mass (*viz.*, $\pi^0 \to \gamma\gamma$, $H^0 \to \gamma\gamma$, $Z^0 \to e^+e^-$), and at the same time to maintain the excellent hadronic energy resolution attained by the fiber-only calorimeter.

The Crystal Calorimeter

The crystal calorimeter is just outside the tracking chamber and consists of $25\,X_0$ of $(2\,\text{cm})^2$ laterally segmented BGO crystals.[62] The physics driving the BGO calorimeter, beyond the precision four-vector measurements mentioned above, is to reconstruct specific final states that carry unique information, such as τ decays into a spin-1 ρ:

$$\tau^{\pm} \to \rho^{\pm}\nu \to \pi^{\pm}\pi^{0}\nu \to \pi^{\pm}\gamma\gamma\nu\,,$$

so that the calorimeter must measure two photons spatially separated by $\Delta(R\phi) \approx R m_{\pi^0}/P_{\pi^0} \approx 2\,\text{cm}$ decaying from a π^0 at $P \approx 10\,\text{GeV/c}$, which decayed from a $\tau^{\pm}$ at 40 GeV. This necessitates $(2\,\text{cm})^2$, or even $(1\,\text{cm})^2$, lateral segmentation.

A simulated τ^- decay is shown in Figure 4.9, where the charged π^- from the vertex is shown impacting the calorimeter very close to the two γs from the π^0 decay, penetrating (without apparent interaction) the BGO and depositing its energy in the fiber calorimeter. The scintillation and Čerenkov signals in the BGO in Figure 4.10b and c display two obvious γs: they are electromagnetic because $C \approx S$ and neutral because there are no charged tracks pointing to the isolated peaks in the BGO. The scintillation and Čerenkov signals in the fiber calorimeter behind the BGO show a single broad structure: it is hadronic because (Chapter 4) Čerenkov and scintillation signals are not equal, the $(C - S)$ difference fluctuates from channel to channel, and it has the shape of a single hadronic particle.

The simplicity (and strength) of dual readout in BGO is illustrated in Figure 7.5a in which the separate Čerenkov and scintillation signals from a single cosmic muon are shown [113] from a digital oscilloscope. The Čerenkov side had a "black

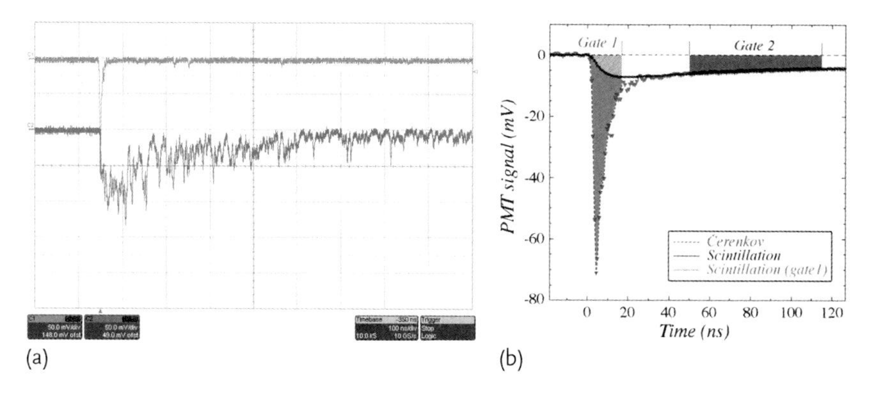

Figure 7.5 (a) 100 ns/div single cosmic μ; (b) averaged pulse showing Čerenkov gate and scintillation gate. (Data from [55]). Panel (a) was obtained by A. Cardini with a single cosmic μ, and (b) is from DREAM beam test data.

62) The crystals tested in the experiment shown in Figure 7.4 were borrowed from the L3 collaboration. We can substitute several crystals for BGO, one being bismuth silicate at substantially less cost and comparable dual-readout performance: "Development of BSO ($Bi_4Si_3O_4$) crystal for radiation detection"[121]. BGO is beam tested by DREAM and thoroughly understood, and therefore our present designs and simulations are for BGO crystals.

filter" passing only UV light, and the scintillation side had a yellow filter appropriate for the emission spectrum of BGO. However, it is clear that two separate readouts are not required[63] and that a single readout with two timing gates, during which times the pulse is summed for scintillation light and Čerenkov light, will also accomplish dual readout of BGO. This is shown in Figure 7.5b on the single digital oscilloscope pulse.

The Fiber Calorimeter

The fiber calorimeter behind the BGO is a spatially fine-grained dual-readout fiber calorimeter with time-history readout at approx. 1 GHz to measure the neutron content of a shower, make other particle identifications, and correct for small attenuation in the scintillating fibers. The calorimeter modules have fibers up to their edges, are constructed for submillimeter close packing, and are projective to the origin with signal extraction on the outside so that the calorimeter system will approach full hermetic coverage without cracks. The total calorimeter depth approaches $10\lambda_{\text{int}}$. The energy resolution calculated in ILCroot is

$$\sigma_E/E \approx 29\%/\sqrt{E} \oplus 1.2\% ,$$

including jet reconstruction in the detector [114], which degrades the single-particle full-calorimeter resolution from the mid-20% range to this 29% statistical term. The 1.2% constant term probably derives from the slight volume nonuniformity of the fiber-to-absorber density at the boundaries of the modules.

The dual-readout principle is to measure the electromagnetic fraction (f_{em}) in each event and directly correct for the different responses event by event. This is illustrated with data in Figure 7.6 showing the Čerenkov and scintillation responses for 100-GeV π^-. For the DREAM module, the separate average Čerenkov and scintillation responses are $(e/h)_C \approx 4.7$ and $(e/h)_S \approx 1.3$ and, consequently, the Čerenkov signal (Q) and the scintillation signal (S) are given by Eqs. (1) and (2) in

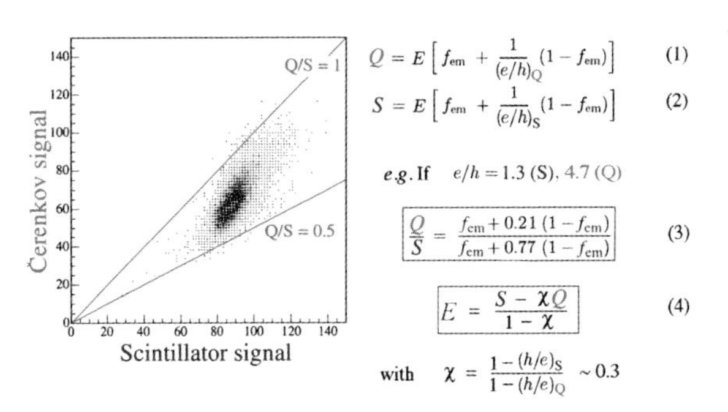

Figure 7.6 The most direct dual-readout plot of Čerenkov signal vs. scintillation signal. On the right are shown the simple linear relationships we use to analyze these data, describing a hadronic shower having two parts: an EM part and a non-EM part, separately weighted by their respective $(e/h)_S$ and $(e/h)_C$ values. (Data from [52]).

63) R. Wigmans.

Figure 7.6. This is the simplest formulation of hadronic calorimeter response: an EM part with response of unity, and a non-EM part with response (h/e). There are two unknowns in each shower, E and f_{em}, and two measurements S and Q. The electromagnetic fraction, f_{em}, is determined entirely by the ratio Q/S, Eq. (3), and the shower energy calculated as in Eq. (4).

The fiber DREAM module has been extensively tested (in a number of different ways), but for our purposes here we show in Figure 7.7a the Čerenkov response to 100-GeV π^- that is characteristic of a noncompensating sampling calorimeter. The response is broad and asymmetric and gets the wrong energy. All of these problems arise from the fact that the hadronic production of π^0s and $\pi^{\pm}$s fluctuates from shower to shower. Since $\pi^0 \rightarrow \gamma\gamma$ exclusively, and the average response of the calorimeter is different for EM particles and hadronic particles $(e/h \neq 1)$, these fluctuations lead to a degradation in energy response. The different response for small slices in f_{em} is shown in Figure 7.7b. Clearly, the broad asymmetric Čerenkov response is just a sum of narrower Gaussian response functions for each f_{em}. This is the key to dual readout.

The Čerenkov response for 200-GeV π^- is shown in the sequence Figure 7.7c → d → e. The raw response of a noncompensating single-readout calorimeter is Figure 7.7c. Using Eq. (4) in Figure 7.6, this dual-readout shower energy, E, is plotted in Figure 7.7d. Immediately, we recover a Gaussian response that is narrower by a factor of 3 and approximately at the correct energy. The DREAM module is $10\lambda_{int}$ deep but only $\sim 1.4\lambda_{int}$ in radius, and therefore there are substantial lateral leakage fluctuations of about $\sigma_{leakage} \approx 4\%$. We can suppress these leakage fluctuations by calculating f_{em} in Eq. (3) as $f_{em} \approx Q/E_{beam}$ instead of $f_{em} \approx Q/E_{shower}$. Fluctuations in S and Q are still present in f_{em}, but effectively the leakage term has been set to zero. The dual-readout resolution with leakage suppressed is shown in Figure 7.7e. The resolutions in Figure 7.7d and e scale in energy as $1/\sqrt{E}$. Reality for a large fiber dual readout of the DREAM geometry is between these two curves.

In addition to this raw energy resolution, the hadronic response of this dual-readout calorimeter is demonstrated to be linear in hadronic energy from 20 to 300 GeV having been *calibrated with 40-GeV electrons* [52] and that this critical linearity will be important at a collider in which a detector can only be calibrated with 45-GeV objects from Z^0 decay but must maintain a true energy up to ten times this energy to do good physics.

The classical compensation mechanism [51] depends on the measurement of the MeV neutrons liberated in the breakup of nuclei, and this has been recently demonstrated in data and simulation by the GLD collaboration [65]. It should be mentioned that one of the original goals of dual-readout calorimetry was to achieve compensation without the limitations on sampling fractions [57]. The resolution improvements achieved by measuring the EM fraction of each event make obvious the next largest fluctuation in hadronic showers, the event-to-event fluctuations in the binding energy losses as hadronic particles of the shower break up nuclei. These binding energy losses are correlated with the liberation of MeV-energy neutrons (much like the spallation protons, but not accelerated away from the nucleus).

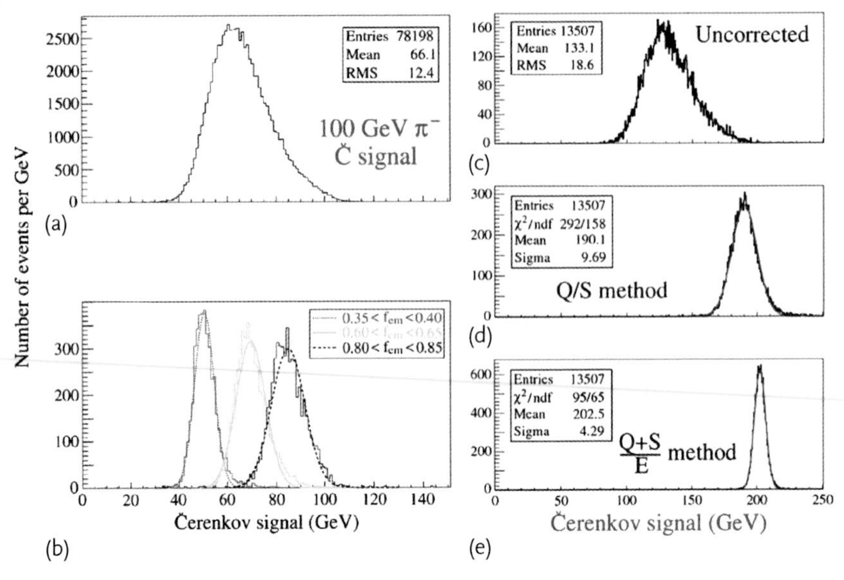

Figure 7.7 The energy resolution of a dual-readout calorimeter is potentially excellent due to measurement of the EM fraction in each hadronic shower. (a) The asymmetric and wide distribution of the Čerenkov signal for 100-GeV π^-; (b) individual slices of f_{em} show clearly that this distribution is a sum of many much narrower distributions; (c) the distribution of the Čerenkov signal for 200-GeV hadrons. It is asymmetric, broad, and not centered at the beam energy; (d) the dual-readout solution to E using Eqs. (1) and (2) of Figure 7.6; and (e) the same solution for E, but with suppression of the leakage fluctuations by calculating f_{em} as (C/E_{beam}), instead of (C/E_{shower}), and then using Eqs. (1) and (2). Note that fluctuations in S and the large statistical photoelectron (pe)-dominated fluctuations in C are still present in f_{em}. (Data from [52]).

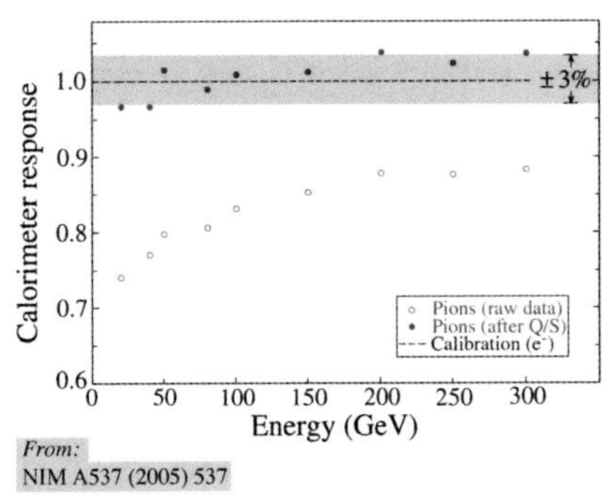

Figure 7.8 Linearity of the hadronic response of the fiber DREAM module from 20–300 GeV. Open circles are for single readout; solid circles are for dual readout. (Data from [52]).

We measure these neutrons as they elastically scatter from protons in the scintillating fibers, $np \rightarrow np$, the recoil proton ionization being the measured signal. The mean neutron energy loss per scatter is $E_n/2$, and therefore the neutrons are quickly moderated to keV energies. We measure this neutron content of a shower by the time history of the scintillation signal since the MeV kinetic energy neutrons travel slowly in the medium, $v_n \sim 0.05c$, and fill a larger volume of the calorimeter.

The following data from DREAM will illustrate this. The neutron fraction, f_n, of a hadronic shower can be measured by summing the scintillation signal from 20 to 40 ns [56] and dividing by the measured (dual-readout) shower energy. The total Čerenkov signal from the DREAM module is plotted against this f_n in Figure 7.9a showing the expected anticorrelation between hadronic energy and electromagnetic energy. This same total Čerenkov signal distribution is shown projected on the x-axis in Figure 7.9b, and it is broad and asymmetric. Taking slices in f_n, we see that

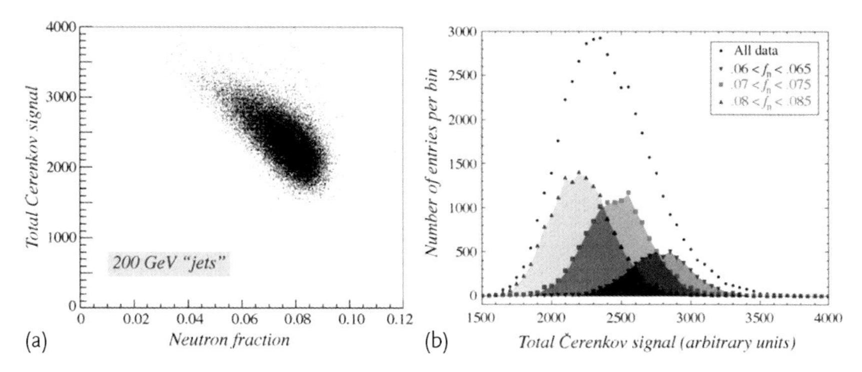

Figure 7.9 (a) The Čerenkov signal plotted against the neutron fraction, f_n, showing the expected anticorrelation between EM and hadronic contents in hadronic showers; (b) the total Čerenkov distribution of (a) projected on the vertical axis and, in addition, the separate distributions of Čerenkov signal for three slices in the neutron fraction, f_n. This is the analogous plot to Figure 7.7 for the electromagnetic fraction, f_{em}, and represents the essential proof that the measured neutron fraction can improve the energy response. (Data from [56]).

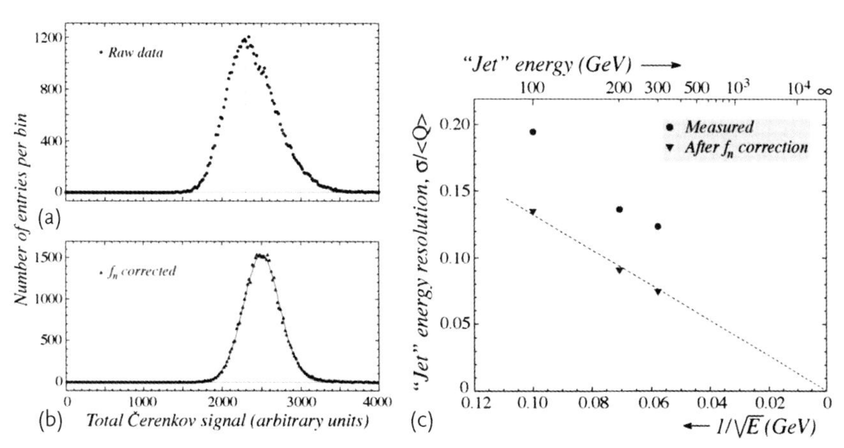

Figure 7.10 (a) The Čerenkov response of DREAM; (b) the Čerenkov response after correction event for event by the neutron fraction, f_n; and (c) the corresponding improvement in the resolution ($\sigma_Q/\bar{Q}$) as a function of hadron energy. (Data from [56]).

this broad and asymmetric total distribution is a sum of narrower Gaussian distributions, similar to what we found with f_{em}, Figure 7.7b.

This measurement of f_n can be used to improve the energy resolution. The raw Čerenkov distribution is shown (again) in Figure 7.10a. Correcting for f_n yields the narrower distribution in Figure 7.10b. The energy dependence of the resolution (rms/mean) for the raw and corrected distributions in Figure 7.10a and b reveals that they both scale with $1/\sqrt{E}$, and that the f_n-corrected resolution does not have an appreciable constant term, as shown in Figure 7.10c.

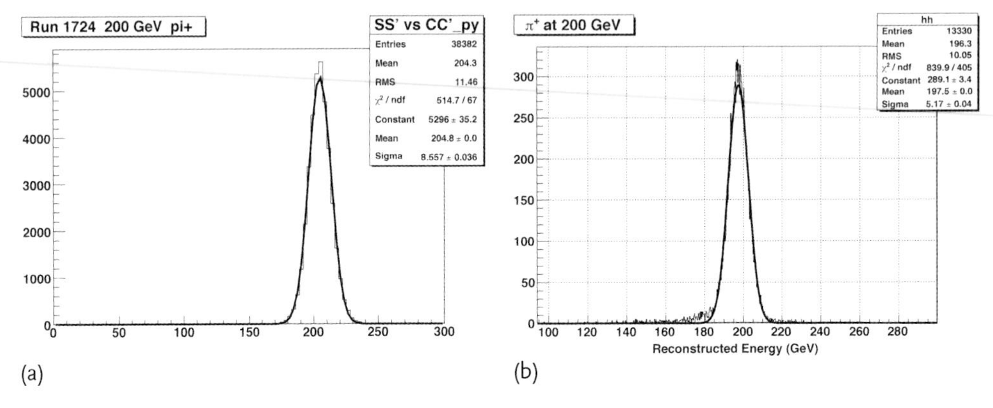

Figure 7.11 (a) The distribution of the reconstructed energy from the DREAM module (a) for a 200-GeV π^+ and from the 4th calorimeter simulated in ILCroot with Fluka (b). *NB:* the data are "cleaner" than the simulation, a reversal of the usual data-Monte Carlo comparison, because the DREAM module is simple (fiber density is constant) whereas the simulation is exceedingly detailed to the level of individual fibers. We actually see a small constant term in the simulation due to fiber placement at the edges of the modules.

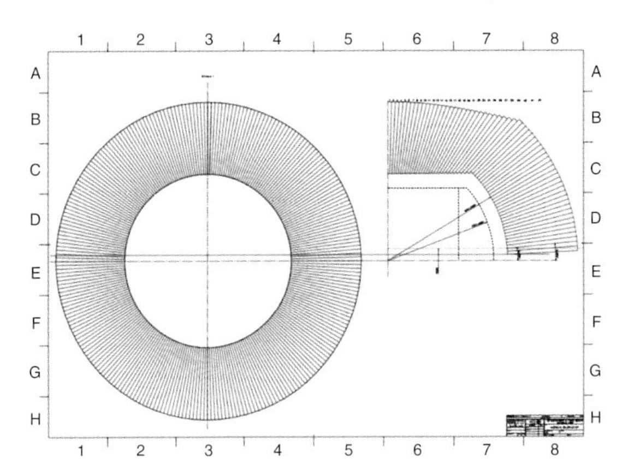

Figure 7.12 Azimuthal segmentation of the hadronic calorimeter at $z = 0$. There are 256 towers in each of the 32 slices in θ. At left is the $r-z$ projection, where one can see the segmentation of 32 concentric tower arrangements of the end cap. In black are the contours of the CluCou chamber. The space between the chamber and the fiber calorimeter is filled with the crystal (EM) calorimeter.

An analysis of the simulated jets also shows that using the neutron measurement improves the energy resolution of jets from $\sim 34\%/\sqrt{E}$ down to $\sim 29\%/\sqrt{E}$. A full discussion of this calorimetry result is given in the talk by Di Benedetto [115] for the projective geometry shown in Figure 7.12.

Finally, a direct comparison of the energy resolution for 200–GeV π^+ is shown in Figure 7.11 for DREAM data (Figure 7.11a) and for 4th simulation (Figure 7.11b).

7.1.4
Magnetic Field Configuration and the Muon System

The field pattern of the dual solenoids is shown in Figure 7.13. A more detailed discussion of the magnetic design is given in the note "Solenoid Design", A. Mikhailichenko (Appendix A of the LoI [75]). The inner solenoid has $\sim$ 33 MA turns and the outer solenoid has $\sim$ 16 MA turns. The coil at the end of the inner solenoid acts as a Helmholtz coil and has $\sim$ 8 MA turns and resides in the same cryovolume as the main coil. Thus, the outer solenoid is easier except for its size. Rutherford-type cable carrying 20 kA is used in both solenoids whose short sample critical current in a helium bath is $\sim$ 40 kA for a safety margin of 2.0.

The muon system consists of essentially a spectrometer between the dual solenoids whose magnetic field configuration is shown in Figure 7.13. This field configuration is easy to obtain and provides a uniform tracking field, a second reverse bending of muons that penetrate the calorimeter, a radial component of

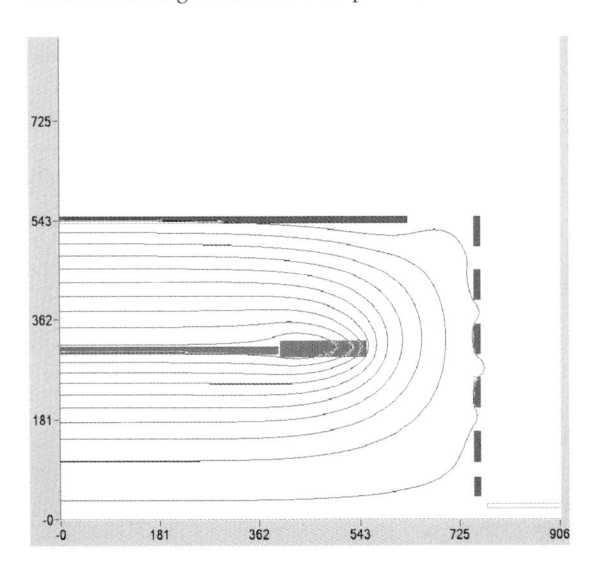

Figure 7.13 Magnetic field configuration of the 4th detector dual solenoids. The central tracking field is $B_z = 3.5$ T, the reverse field between the solenoids is $B_z = -1.5$ T, and the field is symmetric in ϕ and smooth everywhere in θ. The field can be exactly reversed to cancel detector asymmetries in high precision quark asymmetry measurements, and the fringe field is essentially zero.

B to bend very low angle muons, a flux return determined entirely by conductors, and zero fringe field.

The CMS-style conductors allow for a relatively compact solenoid. The field quality is excellent as shown in Figure 7.15, in which each gradation is $\Delta B/B \approx 6{\cdot}10^{-5}$, leading to 10^{-3} variation inside ± 1.5 m and $< 1.4 \cdot 10^{-4}$ inside ± 1.2 m. The field uniformity is shown for slightly elongated solenoids in Figure 6.8, with completely safe peak ***B*** fields inside the superconducting cables, as shown in a blowup of the region of the Helmholtz coils at the ends of the inner solenoid in Figure 7.14. A view of the dual solenoids in their cyrostats in the IR is shown in Figure 7.14.

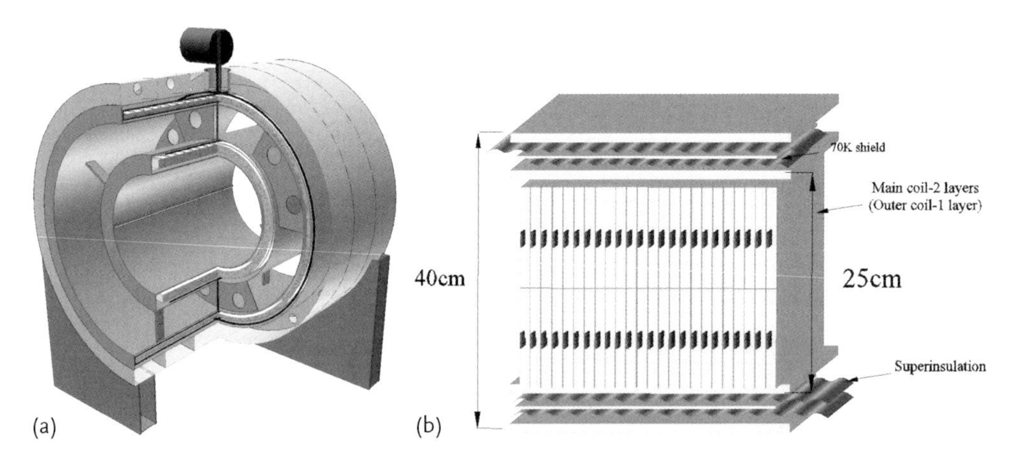

Figure 7.14 (a) View of the two solenoids and their supports. (b) CMS style conductors designed for the 4th inner solenoid.

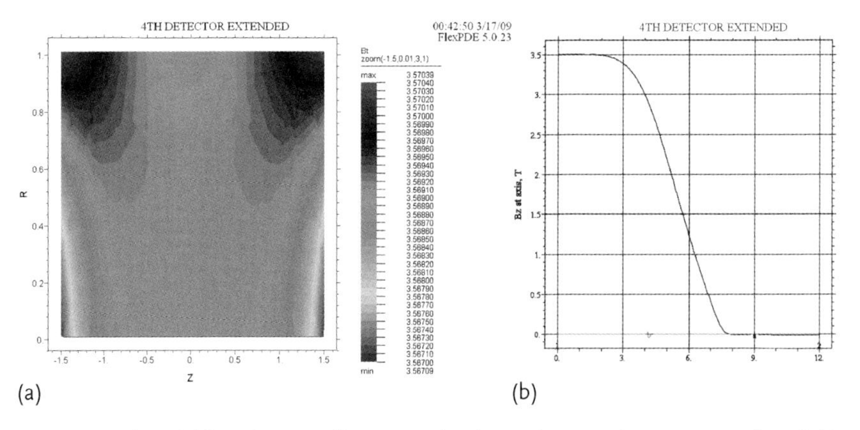

Figure 7.15 The ***B*** field quality is excellent: (a) each color gradation is $\Delta B/B \approx 6 \cdot 10^{-5}$; and (b) the main axial field B_z goes to zero at 7.5 m, thereby protecting the final focus elements.

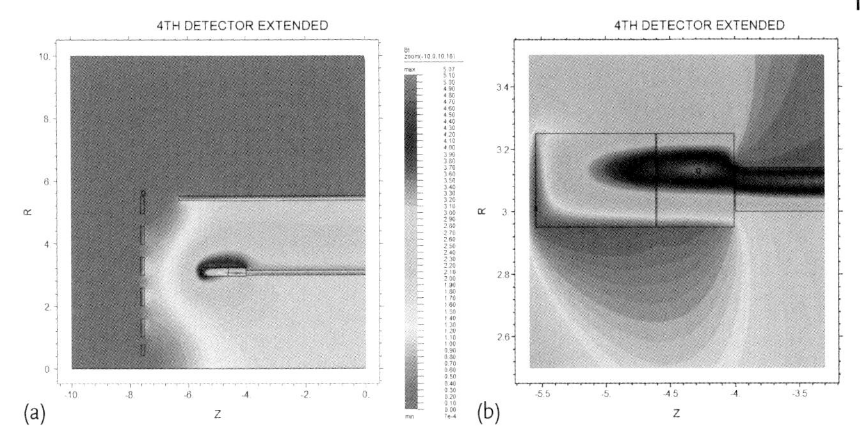

Figure 7.16 (a) The spatial uniformity of the ***B*** field is excellent, and very acceptable for the tracking system. The fringe field is essentially zero; (b) detail of the field intensity in the Helmholtz ends, showing a completely acceptable peak field of $B = 5.08\,\mathrm{T}$.

This open magnetic geometry also allows the following in a detector:

1. final focus (FF) elements can be inside the detector for better control of the beams;
2. a future $\gamma\gamma$-collider option that requires access to the detector;
3. the dual solenoids allow a reversal of the ***B*** field everywhere, and therefore a cancellation of detector-induced asymmetries in the measurements of c, b quark asymmetries. This also applies to $\tau^+\tau^-$ polarization measurements. It is now possible to consider e^+ beams with 80% polarization [116, 117], which in turns allows for physics with polarization amplitudes;
4. the dual solenoids allow for running at a reduced ***B*** field. A zero-field detector has been advocated and implemented by several groups, notably D0 at Fermilab and UA2 at CERN, on the grounds that a magnetic field spreads the charged particles in ϕ, making jet reconstruction more difficult. Others argue that a weak field $B_z \sim 1\,\mathrm{T}$ to measure the sign of $e^\pm, \mu^\pm$ is sufficient for good physics and to thereafter rely on calorimetry for high precision four-vectors. The 4th design allows for running at any field[64] up to the maximum of $B_z = 3.5\,\mathrm{T}$.
5. the open geometry between the solenoids allows future insertions of specialized detectors, for example, to reconstruct massive particles with lifetimes in the millisecond range;
6. the overall detector is lightweight and movable and can be easily disassembled and reconfigured.

Finally, every subsystem makes measurements in a spatial volume (e.g., a calorimeter volume or a tracking channel volume) that are digitized at 1 GHz or faster

64) The CluCou time-distance relationship changes only a little, and in a predictable way, for lower fields.

for the entire interbunch period of 337 ns (for ILC), and therefore *there is never a nanosecond during which nearly every cubic centimeter of the detector is not being watched.* This has proven powerful in dual readout and in cluster counting, but we believe it will also be a benefit in assessing machine backgrounds, "flyers", stray beam, albedo, and so on, during physics running. It will also be useful in the event that massive slow particles are produced, or particles with long mean lifetimes ($\tau \sim \mu s \rightarrow ms$) whose decay products pass through the tracking system and the calorimeter. A detector with continuous time-history readout will be able to see these odd things, as if in a slow-motion movie.

7.2
Dual-solenoid Detectors and Machine–Detector Interface (mdi)

The mdi is complex for future colliders, primarily because the incoming beams are so small (vertically, 4 nm for ILC, and 1 nm for CLIC) that the FF elements that would be, in principle, part of the collider must now also be part of the detector. There are many more issues, such as movement of the detectors in the IR, floor loading, and so on, that do not relate directly to detectors. The gross dimensions of the IR are not critical for a dual-solenoid detector since, it turns out, both the radial and axial dimensions are similar, even smaller, than for an iron-based detector. There are consequences for both the machine and the detector [109, 118].

7.2.1
Detector Benefits

Strict confinement of the magnetic field inside the cylindrical volume defined by the outer solenoid and the "end coils" has several beneficial consequences, as follows.

1. It avoids the huge ($\sim$ 25 000 t) internal forces on the iron volumes when a solenoidal field is energized. These forces in iron-based detectors are toward the center of the detector (the field-generating solenoid) since the energy of the system is lowered when iron of permeability $\mu \gg \mu_0$ is moved into the field region.[65] The two solenoids are about in equilibrium with a small radial centering force of 0.8 t/mm and an even smaller axial decentering force of 0.4 t/mm for the field configuration in Figure 7.13.
2. It avoids the CMS problem of providing enough iron in the pole-tip regions to capture all the 4-T flux from the solenoids before the iron saturates at $B_{\mathrm{Fe}}^{\mathrm{max}} \approx$ 1.8 T. This problem is exacerbated at higher B, and some problems have been encountered in recent calculations of fringe fields in linear collider detectors.

65) These forces move parts of CMS (near HF) axially by as much as 19 mm, a spatial distortion that is too large for a precision linear collider detector.

Adding more iron solves the flux density problem but does not help muon identification and results in a more massive detector.

3. All regions of the field volume are useful for physics: the uniform central tracking field, the nearly uniform annulus region between the solenoids, and the very forward region for small angle muon momentum measurement are available for sagitta measurements.

The no-iron detector is easily disassembled into its modular components and reassembled. This will be important for modifications or additions to the detector over a long lifetime.

1. One seldom contemplates modifications to a large detector; however, the Lead Glass Wall was added to the Mark II, and it is conceivable that completely new objects might exist (e.g., an LSP with high mass, low velocity, or a long lifetime) that would penetrate the calorimeter and enter the muon tracking region, which could be reinstrumented to enhance the efficiency for seeing such an object.
2. Quick repairs (time scale of 1 month) are easy in this open geometry, in contrast to an iron-based system with its weight and sarcophagus-like geometrical restrictions.

The absence of the iron-mass allows a lightweight and open detector with flexible functionality, and open geometry so that

1. surveying and alignments and realignments are easier, and this complements the precision of the individual detector components, which require the global coalignments of all detector subsystems, in particular the vertex chamber;
2. all elements of the final focus are visible from a single point. Therefore, a laser interferometric alignment system can be used to link external points to critical detector locations.

7.2.2 Machine and Interaction Region (IR) Benefits

Strict confinement of the magnetic field has many implications for the machine.

1. The most important is the absence of a fringe field in the IR, which in turn has many secondary implications in small forces and small effects on the high precision machine components.
2. The final focus magnetic elements need not be "shielded" from the stray field, a measure that in itself is problematic.

3. All magnetic elements (quadrupoles, sextupoles, kickers, etc.) of the FF are visible from a single point for direct alignment among themselves, and also critical detector alignment with respect to the beam.
4. Strict confinement of the magnetic field allows for easy incorporation of a laser optical system for $\gamma\gamma$ collisions, as recently suggested by Dr. Sugawara as a prelinear collider, as proposed for many years by the Budker Institute [119] and by M. Peskin and the Physics Common Task group.

The absence of the iron mass has implications for mechanical and radiation dose issues.

1. Almost every problem in the IR becomes easier without the 15 000-tonne iron mass, from crane capacities to platform deformations and floor loading, and considering the general mobility required for any detector shuffling schemes.
2. The iron mass is not available for the gravitational support of the detector components.
3. The iron mass provides for "self-shielding" of the detector in the IR.Eqs.[66] The dose at an electron machine is very low compared to the nuclear dose at a proton machine [18], and it may be that a dual solenoid is not a good idea at a proton machine. In this respect, the experience of ATLAS will be important [74].

7.2.3 Physics Benefits

1. The magnetic field of the experiment can be reversed, $\boldsymbol{B} \to -\boldsymbol{B}$, to cancel detector-induced asymmetries in physics polarization measurements [116] and any other asymmetries that depend on detector geometry and field configuration. The D0 experiment at the Tevatron Collider routinely reverses both the 2-T solenoid tracking field and the iron-based toroidal field for just these reasons.
2. The experiment can be run at reduced field, for example, $B \approx 1\,\text{T}$, or even $B = 0$, for special runs or to test some physics with a calorimeter-only detector. To my knowledge, this has never been done in an experiment.
3. Precision muon tracking in the annulus between the solenoids allows for a complete tracking image of everything that exits the $10\lambda_{\text{int}}$ calorimeter and the solenoid, with both exploratory physics and muon identification possibilities.
4. The alignment and survey benefits mentioned above impact the quality of physics through calibrations and stability of the detector.

66) The radiation requirements for the ILC IR are based on wide laboratory practice, and initial calculations by A. Seryi and T. Sanami [19] indicate that 2.5 m of concrete shielding is sufficient to satisfy the radiation requirements for the maximal accident: a full train (2820 bunches) at 250 GeV hitting a 25-cm Cu plug anywhere along the beam inside the detector to maintain a dose of less than 250 mSv/h, or an integrated dose of 1 mSv per accident, anywhere in the IR.

Final Comment on Dual Solenoids the fringe field is small and can be made zero,[67] even near the detector, since the field configuration is defined by currents and not by iron. Fundamentally, the magnetic field is simple because it is natural, for example, there are no sharp turns, no saturated iron masses, no discontinuities between permeable and nonpermeable materials, and no iron plates separated by air gaps and terminating in air at the extremities of the detector. All of these problems are avoided by returning the flux with another solenoid, rather than iron.

7.3 Problems

1. Estimate the mass resolution on a $Z' \to q\bar{q}$ of mass 1 TeV/c^2 for a hadronic calorimeter with energy resolution implied by Figure 7.11b. Assume the Z' is produced near rest in the detector, $e^+e^- \to Z' \to j\,j$. Consider calorimeter constant terms of 0.1%, 1%, and 3%.

2. Generally, how well will this Z' in Problem 1 be measured in the decay modes $Z' \to e^+e^-$ and $Z' \to \mu^+\mu^-$?

3. It has been claimed that the centering (or decentering) forces between the solenoids of a dual-solenoid system are small, only a fraction of a ton per millimeter of axial or radial displacement. Justify or deny this claim. Consider the value of B^2 and how it changes as one solenoid is moved relative to the other.

4. Quench protection is important for large superconducting magnets and solenoids. For dual solenoids, there are two, and if one solenoid quenches and the other does not, a huge magnetic dipole fringe field develops (over tens of seconds). How would you protect against this happening?

5. If dual readout is good, would any form of triple or quadruple readout be better? Familiarize yourself with hadronic interactions (Wigmans [29] or another source), list as many physically distinct populations of particles ($e^{\pm}, n, p, \ldots$) as you can, and devise a means of measuring each. It doesn't matter that the means may be unrealistic, at the moment. The main question is: after measuring the electromagnetic and neutron fractions, f_{em} and f_n, are there further fluctuations that exist to be measured?

6. The first criticism of the dual solenoid was that you do not need to measure the muon momentum a second time (in the annulus between the solenoids) since it is already measured to high accuracy in the main tracking chamber. You merely need to properly associate a track emerging from the calorimeter solenoid with the proper track in the tracking chamber. This is a perfectly correct argument. Estimate the misassociation probability for a muon next to a jet in an event, and either look up, use GEANT, or guess at the particle punch-

67) An expansion of the field in multipole components and direct trim coil cancellations are one means.

through rate into the region after the solenoid. (Of course, this depends on the depth of the calorimeter.)

7. The dual-solenoid solution requires a very large outer solenoid. This problem explores the optimum radius of this outer solenoid, given a fixed size inner solenoid. Assume that the radius, current, and turn density in the inner solenoid are fixed at $R_1 = 3.3\,\text{m}$, $i_1 = 20\,\text{kA}$, and $n_1 = 200$ turns/m. In all calculations, make it simple by calculating only the pure solenoid field as $B = \mu_0 n i$, considering only the field within the solenoid, and ignoring the end field that changes from B_1 to B_2 (this is just an average of the two uniform fields in the two solenoids). The only independent parameter is R_2, the radius of the outer solenoid, and the problems (a–g) should be done as a function of R_2 varying from 4 to 10 m. The ampere turns in the outer solenoid are determined by the requirement that the flux of the inner solenoid must fill the annulus between the solenoids. From this, everything can be calculated.

   a) What is the flux density, or field B_1, inside the inner solenoid? (Note that B_1 will depend on R_2.)
   b) What is the flux density, or field B_2, in the annulus between the solenoids?
   c) What is the magnetic stored energy in the system?
   d) How does the momentum resolution in the main tracker (assume fixed track lengths, $\ell = 1\,\text{m}$, inside the B_1 field) change with R_2?
   e) How does the momentum resolution of a track between the annulus change with R_2, assuming that the track length is $\ell \approx R_2 - R_1$?
   f) What is the "magnetic pressure" on the outer solenoid?
   g) Consider the possibility of a water-cooled Cu outer solenoid. What is the power required in this solenoid, that is, what are the Joule $I^2 R$ losses? The resistivity of Cu is $\rho = 1.7 \times 10^{-8}\,\Omega \cdot \text{m}$. For each R_2, you might want to optimize the geometry of the Cu conductors to minimize $I^2 R$.

8. It would be a good idea to pose a comparable question about optimizing an iron-based flux return, but the calculations cannot be so easily performed, even approximately, due to the permeability of iron and the relaxation-type numerical codes that are required to find an accurate solution. The complexity is indicated in Figure 6.15 showing the flux lines and (in color in the original plot) the color-coded flux densities. Since the magnetic energy goes to zero (approximately) for a flux line inside iron, the entire flux stays inside an iron mass until it must cross an air gap into another iron mass. That is all that is happening in this plot, and the many pieces of iron, including the iron in the floor and the elevator shafts, make it a complicated problem. Nevertheless, from Figure 6.15 try to guess where you would put more iron to confine the errant field lines.

8
The Sociology of Collaborations

"You're going to get a degree in psychology, too."
– Professor to his student

8.1
Big vs. Small

It is clear that, in the end, only the big labs CERN, Fermilab, KEK, and SLAC have the resources in terms of stable funding, engineering skills, management oversight, and so on, to build the big detectors at costs of \$0.3–0.5 billion apiece with construction schedules over periods of 10 years or more. This is a very complex operation requiring skills we physicists never learned in graduate school, skills that combine management engineering, technical knowledge, social psychology, financial acumen, and the critical ability to keep people working toward a common goal while simultaneously ensuring individual rewards and commensurate accountability. To make this problem even more difficult, each group that contributes to a detector system must be able to go back to its funding agency and honestly say that a particular piece of this big detector belongs to them, that it is their "territory." Therefore, in the largest of detectors built to date, a tracking system might be divided up into several pieces, for example, inner layers, middle layers, outer layers, and end-cap disks, with each piece of real estate given to a separate group. This is a technical and political imperative. It can lead to exceedingly awkward and challenging engineering problems, as can be imagined, where one piece of a detector must support another piece, but the dimensions are not quite right. The World Wide Web was invented by Timothy Berners-Lee [122] to solve exactly this problem: the coordination of technical and scientific work among groups and individuals at distant locations. However, without good planning (and even with good planning, but without sufficient attention to detail) it is the "end game" of construction that is dangerous. There are funds from many sources, pressures from schedules, and technical challenges that much be solved day by day.

In the early stages of a project, the situation is quite different, partly because there is much less at stake in the funding decisions to be made, but primarily because new ideas have a chance to surface and, in the fortunate experiments,

Particle Physics Experiments at High Energy Colliders. John Hauptman

ISBN: 978-3-527-40825-2

there is enough time to weed out the bad and unworkable ideas and to pursue the good and tested ideas.

I heard of a case where a big lab was going to buy a huge number of heavy scintillating crystals to fashion into a cubic-meter hadronic calorimeter and to test it in a beam. This would involve hundreds of high voltage channels, many hundreds of photoconverters (e.g., SiPMs), an extensive data-acquisition system, and concerns about temperature stability, geometry nonuniformities, and so on. In my judgment, this would have resulted in a large and expensive failure in the beam.

In contrast, a small group of individual university physicists were addressing the same problem but, able only to borrow one crystal and using a digital oscilloscope as a readout platform, performed a thorough testing using single particles (e, μ, π) as probes, different PMTs on each end of the crystal, various optical filters between the crystal and the PMTs, over a 40°C temperature range, and mounted on a simple rotating platform to measure the angular dependence of the scintillations and Čerenkov components of the light. An axial scan of the crystal was also made to measure directly the attenuation of both scintillation light and Čerenkov light toward both ends of the crystal [53, 54]. These data provide a complete characterization of the crystal, and on this basis, several crystals were rejected as unworkable for a calorimeter.

Therefore, big labs have the weakness that their physicists "think big" and with their available resources jump quickly into a large problem. University physicists with far fewer resources "think small" and develop a fundamental foundation from which a scaling-up can be performed with technical confidence. Both talents are needed, but in the correct order.

An experienced physicist might say that all of this is true, but we have a schedule to meet and a fixed budget, we do not have the time or luxury of a first-principles R&D program, we have to go with what we've got right now. This is also true, in many circumstances. There are famous counterexamples to this big- vs. small-difference, and one is Georges Charpak at CERN, who developed multiwire proportional chambers and drift chambers on a small scale, clearly pursuing the problem from first principles. It is more difficult to find a counterexample in the other direction.

8.2 Structure of a Group

Consider a mythical small group, maybe just two or three people, who are thinking of an experiment, maybe at a future collider but not necessarily. Usually, there is a central theme, idea, or technology that shapes the early ideas, but in any case this small group begins to formulate ideas in a very general way, sometimes borrowing solutions from previous experiments as safety blankets. The group enlarges the circle of discussions, much like Steinberger did in the pre-ALEPH group, in his words (Section 6.2), or as Nygren did in the pre-TPC collaboration,[68] which is particularly easy at a major laboratory with its many interested and capable people. This early thinking is sometimes based on beam tests, sometimes on bench tests of new technologies, or sometimes on the collective experience of the participants. However, a credible detector design will entail the construction and testing of prototypes, funds, and commitments by groups, from which a collaboration is formed, which may have begun as a protocollaboration. At this stage, the most important decisions are made about technologies with options that can be exercised, the degree of engineering of prototypes, a credible simulation of the design, including the choice of collaborators, which may affect later funding prospects.

There are three general dangers in the life of a big experiment and its collaboration, in chronological order of appearance.

1. As the collaboration takes shape, there are competing proposals for the tracking systems and the calorimeters, in which it is generally smart to say "yes" to every proposal, suggesting simulations and beam tests of each, hoping for an early shakeout and, therefore, an easy decision. When there is no clear winner, and a collaboration is under funding and scheduling duress to produce a proposal, a large group with funding can say "we will build the calorimeter", to which the answer must be "yes." This may not be the best scientific choice, but it is the only choice that might allow completion of a proposal or a technical design report (TDR), both of which have deadlines and criteria, and the reputation of the collaboration depends on both.
2. The middle game is the most difficult with its varied and multiple tests of technologies, beam tests that are never perfect, mistakes and misunderstandings of data, and competition for one's own detector with all the glory and funds that come with it. An example of a misreading of data can occur in the all-important hadronic calorimeter beam test with a module that is not large enough. Leakage is apparent and a selection is made of events that do not leak much energy by a number of criteria. This selection, however, enriches the sample in events that are predominantly electromagnetic. After this selection, the energy resolution looks very good (because it is the electromagnetic resolution, not the hadronic), and this is misinterpreted as hadronic energy resolution. Further examples are not hard to guess, but the essential danger of a beam test is that severe selections can be made on the event sample, for example, the impact point of the beam particle on the detector, leaving a biased sample. The rule is that you should not use any information that a particle from a W decay would not know and, furthermore, not use data from the detector itself to make any selections. The critical physicist will always count the number of beam triggers and compare the numbers of events in the final plots with this number.

68) These two groups, the Nygren and Steinberger models, are almost perfect in allowing diversity of opinions in the beginning, keeping physics goals in sight at all times, and forming trust within the protocollaboraton so that difficult decisions can be handled when they arise later. A group that can negotiate these problems can also negotiate good physics.

It is between these two stages, the second and third, that a fundamental change in character should be made: the physicists have to stop having ideas, they have to stop improving things, and let the engineers and the talented builders take over without interference. There is plenty of work still to do, of course, in physics software and in shoring up the group for working together in the long haul of stage three.

3. After the detector is fully designed and its physics potential completely vetted and understood, the construction phase begins with a heavy emphasis on engineering and scheduling of resources. This is best left to real engineers, or physicists with experience in big projects. However, the danger lies in a mismatch between the physics TDR and the engineering necessities of supports, cooling, cabling, and services to the subsystems of the detector. These take the forms of "dead volumes", loss of solid angular coverage, too much material in the tracking system, and too little material in the calorimeters. It happens when another 20% cooling is required, or another 50% in support material, or when two detector systems overlap in space.[69]

A group should evolve in time to suit the needs and character of the work. In the beginning, complete freedom and independence is valuable all around, whereas in the ending stages of a big project an almost completely rigid structure is required to finish the project on time and on schedule. The middle stages are the most challenging in the sense that collaborators are chosen, funds are secured, there is competition for positions and authority, and there can be a very large number of people in 20 to 30 separate subprojects of the overall experiment. These are very different talents, and few, if anyone, among us possess them all. The organization charts in Figure 8.1 are meant to suggest that different structures are appropriate at different stages in a collaboration. On the 4th detector, I avoided ever making an organization chart for two reasons: first, there was too little at stake and, second, a organization chart fixes people into positions and stifles initiative and creativity, exactly the two drivers of innovation and new solutions that one wants at the beginning of an experiment. I was not there, but my guess is that Steinberger's weekly group did not discuss organization charts.

Let me quickly add that in the middle and ending stages, organizational charts (properly done) are essential for accountability and chain-of-command in which projects and subprojects must be completed and decisions made. In fact, it is the funding agency that demands an organization chart: someone has to be responsible for failure.

The costs and scale of high energy detectors has driven physicists into larger and larger groups over the past four decades in which cooperation within the group has advantages, and personal dynamics counts, maybe more than particle dynamics. Let me relate an apocryphal story that came to me while sitting at a long meeting at CERN. "When two men enter the room, they start circling around each other trying

69) The PHENIX detector inserted 2 cm of space between each system during the design, private communication, B. Jacak.

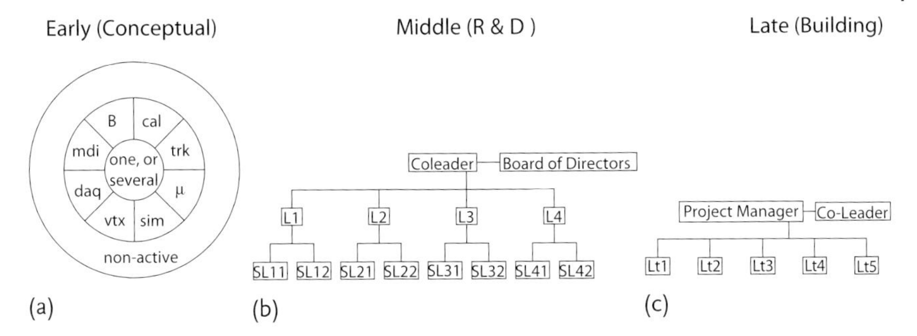

Figure 8.1 Possible organization charts for (a) the early days and months of a group, (b) the middle years of mostly detector R&D, and (c) the later, ending years of construction and scheduling.

to figure out who will be dominant. When a woman enters the room, she looks around and starts thinking how she is going to get these people to work together." It is not at all surprising that women are in leadership positions throughout CERN and Fermilab.

There is a lot written on the workings of large groups and by people for whom it counts, for example, in the business world. The book by Larson and LaFasto [123] points the finger for failure at those people within collaborations that have their own agenda. In a detector collaboration, this can be a person with funding for one component and it is consequently forced on the collaboration, or a person who strives for political control either through personal connections or control of resources.

The Person without a Blemish When an experiment starts to build and test prototypes, and eventually the large detector itself, there are those who come fourth immediately and start working. There would be no experiment without them. There are others who hang back, do not become involved so quickly. Over the course of time, those actually working make a few mistakes, some large and some small, while those not working (but possibly offering free advice to others) are mistake free. I have seen this dynamic several times in experiments, and it reminds me of the *six stages of a project* that Marjorie Shapiro identified during the difficult years of building the TPC:

1. wild enthusiasm,
2. disillusionment,
3. total confusion,
4. search for the guilty,
5. punishment of the innocent,
6. promotion of the nonparticipants.

This may be a truism about human nature. Those who hang back, or join an experiment after all the work is done, capitalize easily on the "physics" and give a lot of talks. Those who worked to build the experiment, and who survive professionally and stay on the experiment, are often thanked by being responsible for the calibration, for the simple reason that they know what they are doing. There must be a way to avoid this problem, maybe by mixing people up among the tasks of a project from the beginning. But it is not so simple. Some hardware experts and builders do not enjoy software, and after years of construction, they are out of touch with the software frameworks and the code, and it would be an exercise in frustration for them to do analysis and give talks. At least the younger people can be mixed up.

8.3 Complexity

I was once asked during a colloquium talk at a university by a non-high-energy physicist whether the big detectors and big colliders had reached a level of complexity that would prevent success. This was during the LHC repairs of 2009. It is a good question, and the complexity is not only technical but, as alluded to above, sociological as well. Is it possible for different groups to be aware of incompatibilities developing in the designs of other groups before installation of their respective equipment?

I suppose there are several answers to this question, and the first ones that come to mind are "Tevatron" and "CDF", "D0", "LHC" and "ATLAS", "CMS", "LEP", and all the successful LEP experiments. This is not just flippant, but actual working detectors are possibly the only answers to an otherwise unanswerable question.

8.4 Software

This book is about the instruments and not the software that simulates those instruments, but it is clear that performance of any system as complex as a big detector cannot be subject to guesswork. In fact, it is common these days (even in a proposal to build and test a small instrument) to ask that a simulation exist to somehow justify the instrument for the purpose of approving a proposal.

Here, software means the simulation of the instruments of the detector and the analysis procedures that use the simulated results to show the overall measurement, or "physics", capabilities of a detector.

Simulations are exceedingly powerful for the understanding of complex systems [124] and have been used for centuries, for example, in the crude calculation of π. For a big detector, there is no alternative to a simulation of all subsystems and detectors, and a framework that allows a single particle to pass from one detector into the next, delivering signals as it goes.

This is fine when a simulation is correct and can actually mimic an instrument, but highly misleading otherwise. This issue is a difficult one when the instruments are hadronic calorimeters since it has been true for many years that some hadronic calorimeter simulations did not properly mimic crucial aspects of hadronic interactions in calorimeter materials, both the absorbers and the sensor.[70] For reasons given in Section 3.3, almost everything is important in a hadronic calorimeter and any superficial or incomplete simulation will lead to incorrect results.

Simulations are also required to calculate the detailed performance of each detector subsystem and to compare these simulations with test beam results, and it is these simulations (when correct) that form the basis, and establish the validity, of the simulation code for the whole detector. This natural tension between simulations and experiments is healthy and only through this procedure, one might even call it the *scientific method,* will reliable and trustable simulation codes be developed. This is not easy, and the plethora of "GEANT lists", each representing a variation in the simulation code, is a measure of this difficulty. Although we, in principle, understand nuclear physics and the properties of nuclei, the violent breakup of a nucleus by a high energy particle that may undergo a single or multiple interactions in a single nucleus, is far from the nuclei in equilibrium that we know from our nuclear physics courses. With respect to hadronic calorimetry and the importance of fluctuations in π^0 production relative to $\pi^{\pm}$ production, that is, the electromagnetic fracion f_{EM}, four measurements [50, 125] in a single beam test can serve to "benchmark" calorimeter simulations, as discussed by Wigmans [29, pp. 523–4]:

Validation of π^0 production in simulation codes

1. the measured nonlinearity of the π/e response ratio vs. particle energy;
2. the measured p/π response ratio;
3. the shape of the response function is Gaussian for incident p and non-Gaussian for incident π; and,
4. the energy resolution differs by 20% (p better than π).

These substantial differences severely test the details of the simulation of π^0s. There is a similar set of measurements to test the correct simulation of neutrons in calorimeters, Wigmans [29, pp. 525–5]:

Validation of slow neutrons in simulation codes

1. energy resolution for electrons and pions vs. beam energy;
2. with varied absorber-hydrogen ratios.

The neutrons provide a fraction of the physical response through elastic neutron scattering off protons, $np \to np$, and it is important to compare, for example, the responses of calorimeters with and without hydrogen (e.g., plastic scintillator) in the calorimeter volume, and to compare both spatial distributions and time distributions for the simulation and the data, since the neutrons are a few MeV in energy, travel at $v \sim 0.05c$, and fill a larger volume than the hadronic shower particles.

Tracking chambers are easier to simulate than calorimeters, and for a superficial or "fast" simulation, just doing a Gaussian smearing of the measurement points

70) See the talk by R. Wigmans [128]

along a track is sufficient. However, for sophisticated tracking chambers such as TPCs with micropatterned end caps [126] and complex silicon pixel or strip systems, a careful attention to electron and hole mobility, impurities, Lorentz forces, and electronics readout should be performed carefully.

Although physicists are educated to maintain generalized capabilities and knowledge, some specialties develop in instrumentation (sometimes called hardware) and computations (sometimes called software). Hardware, beam tests, and instruments are costly and difficult. Software simulation puts it all together, and at this point the software person often takes the mistaken notion that software rules the experiment.

I have chosen to put the topic of software into the sociology section because, with a few pleasant exceptions, the software area often becomes a battleground of different opinions and conflicting egos precisely because it is the point at which the detector comes together as a whole into a single view, where the first "physics" is done in a wholly credible way, and the visibility of a person in control of the software is guaranteed. Often there is a mismatch between an understanding of the detectors and instrumentation and the person writing the software, and so it is the responsibility of the software person to understand the physics of the instruments.

However, the simulation and software are highly critical for a big detector, not because it is always, or even often, correct, but because we have grown to be accustomed to it and to trust it. We believe it even when it is wrong, because there is nothing else to believe. I have worked on physics analysis in big experiments where there are dozens of "correction factors" that are used to correct for effects in the simulation that do not match the data. There are three possibilities: the physics model (or generator) is wrong, the simulation is wrong, or the detector is not what you think it is. Of course, a realist might say that it can be all three at once. One experiment had a "missing radiation length" of material that was not in the simulation code but that was clearly in the detector, as was shown by measuring the $\gamma \rightarrow e^+e^-$ conversion rate in (x, y, z) space. There is no substitute for stringent beam testing of all instruments before the massive construction begins, and the software and simulation should start at this point so that the transition from test beam simulation to 4π detector simulation is not a blind shot. Finally, in the end, sufficiently excellent detectors can use the detector resolutions themselves to interrogate the response of the detector to assess the difficult problems of missing material, biases, and mismeasurements.

8.5 Funding

Some large detector collaborations have learned the trick practiced in US military projects of putting a piece of the project along with jobs and funds into every congressional district. This essentially guarantees passage of the legislation for this project. In a similar way, a big detector project will pass out parts of its detector systems to different institutes and universities, thereby guaranteeing funding for the project. It is obvious that a good detector is not guaranteed by this procedure.

8.6 International and Cultural Aspects

> *"We can't solve problems by using the same kind of thinking we used when we created them."*
>
> – Albert Einstein

In high energy physics, we live and work in a rather unique situation. The big experiments and the big international labs where they reside play a large and important role in international scientific relations. It is more than just collaborators from across the world who visit, work, and live at the big labs, such as we see so clearly at CERN with the LHC and at Fermilab with the Tevatron. It is more than just the many scientific friendships that develop and thrive. It is the creation of a "scientific united nations" within which all scientists are welcome to participate. National interests and conflicts recede and human contacts emerge. Because physicists are often respected and are sometimes influential in their governments, the attitudes and opinions of physicists count.

Herwig Schopper wrote:[71] "To my knowledge, CERN is the only laboratory created to foster science and international collaboration", and one might say that, in the beginning, "international" mostly meant European as it emerged from the economic devastation of World War II. But a good idea cannot be confined, and CERN quickly became a truly international laboratory. The first director-general and a prime mover of international collaboration was Eduardo Amaldo,[72] followed by Victor Weisskopf.

Jean-Pierre Delahaye "I have personally been convinced during my whole professional life of the great advantages and the increasing necessity of International Collaboration. This is one of the main reasons why I work at CERN. As many of you know, CERN was created in September 1954 not only to perform high-energy physics research, but also to allow people from different nationalities who fought each other during the first and second world wars to work together in order that such horrors never happen again. The visionary scientists like Louis de Broglie, Eduardo Amaldi, Lew Kowarski, Niels Bohr and others who promoted the idea, were convinced that by working together, people from different nationalities would not only get to know and appreciate each other better, but they would realise that 'good guys' and 'bad guys' exist in every country and nationality, with a proportion

71) Herwig Schopper, director-general of CERN, 1981–88, in the *CERN Courier*, 29 March 1999.

72) Amaldi's resume is breathtaking in science and scientific leadership: *CERN Courier*, 8 December 2008.

of each corresponding to a constant of mankind. Demonstrating the universality of this constant could constitute an interesting thesis in sociology."[73]

Robert Wilson On the other side of the Atlantic was a nation barely scathed by war that, throughout its large populace, had little interest in international collaboration. Out of the Manhattan Project and its destruction of Hiroshima and Nagasaki came many physicists convinced that "the same kind of thinking" was not working very well, and one of these was Robert R. Wilson, who built Fermilab ahead of schedule and under budget, and doubled its energy with superconducting magnets. He was courageous in the political arena as well, and a hero among young people for this exchange with Senator John Pastore before the Congressional Joint Committee on Atomic Energy for funding of $200 million for the new Fermi National Accelerator Laboratory:

Pastore: *Is there anything connected in the hopes of this accelerator that in any way involves the security of this country?*
Wilson: *No sir; I do not believe so.*
Pastore: *Nothing at all?*
Wilson: *Nothing at all.*
Pastore: *It has no value in that respect?*
Wilson: *It only has to do with the respect with which we regard one another, the dignity of men, our love of culture. It has to do with those things. It has nothing to do with the military, I am sorry.*
Pastore: *Don't be sorry for it.*
Wilson: *I am not, but I cannot in honesty say it has any such application.*
Pastore: *Is there anything here that projects us in a position of being competitive with the Russians, with regard to this race?*
Wilson: *Only from a long-range point of view, of a developing technology. Otherwise, it has to do with: Are we good painters, good sculptors, great poets? I mean all the things that we really venerate and honor in our country and are patriotic about. In that sense, this new knowledge has all to do with honor and country but it has nothing to do directly with defending our country, except to make it worth defending.*

During these years of the Cold War, when our thinking was more or less driven by fear, it was my good fortune to be a postdoc on two experiments[74] at Fermilab, supported by Robert Wilson over the objections of his PAC, to measure the electromagnetic size of the π^- and K^- mesons through direct $\pi^- e^-$ and $K^- e^-$ scattering experiments in a LH_2 target. There was very little good science in these experiments, and at best any results we achieved were too crude to test any model, but we designed, worked, took data, analyzed, and published. We learned many small things, that Russians were more accurate programmers than Americans, that logic chips in Fermilab stores are free for the taking, and that we still had linger-

73) ILCNewsline, 29 August 2009.
74) E216 and E456 in the Meson Lab.

ing suspicions about intentions, but in the end that we did physics together and enjoyed it. The big things we learned are almost too trite to mention: that humans and their aspirations are the same everywhere, that everyone loves their children, and that we who work together and eat together are not enemies.

Andrew Sessler and Edmund Wilson In *Engines of Discovery: A Century of Particle Accelerators*, Sessler and Wilson [6] write:

> "The endeavor of building these machines then brings together young people from many nations. It teaches them that – in a world where reason is the ultimate test of validity for any new idea, and where politics must therefore by subservient – there need be no barrier between nationality, race or creed.
> In the formation of this community of scientists, the prejudices which separated the warring states of Europe were the first to crumble. Those who came to CERN in the 1950s immediately shed the propaganda which had clothed their thinking for a decade. Even during the darker days of the cold war, scientific contacts with Russia were maintained in the field of accelerators and particle physics. As Europe has unified and the iron curtain has fallen, new states have looked to particle physics research for their first tentative step towards a broader political union."

Takeo Kawamura More recently, Takeo Kawamura of the Japanese Diet gave a talk at the TILC08 meeting [127].

> "In the history of civilization, human beings have been nurtured by culture and art, and we are pushing the boundaries of science with the mind challenging the mystery of nature and life. We feel satisfied when the desire to know or learn is fulfilled, feeling a sense of high contentment or spiritual richness. And when we feel spiritual richness, we will have a sense of happiness being a human, making us feel grateful to be born in this world.
> In the Japanese constitution, it declares [our] will to hold an honorable position in international society. I think it is time for us to put our effort into becoming the world's top nation in spiritual richness. And the spiritual richness derived from science must rank with that from culture and art. The nation's maturation is no longer measured only by economic indicators such as GDP. We have new indicators to measure a nation's maturation such as GNH (Gross National Happiness). Now we should regard basic science as one of these indicators."

Where to Put the "World Machine"? Many years ago ICFA (International Committee for Future Accelerators)[75] considered where to site the next big machine, a true world machine. The US and the USSR, and their allies, were eliminated from consideration. The main requirements were (i) an international airport close to the site, (ii) centrally located, (iii) an educated population that would serve all the functions (translations, facilities, cultural amenities) of an international community of physicists and their families, (iv) sufficient national wealth to perform all of the re-

75) Formed by the International Union of Pure and Applied Physics (IUPAP). This story I remember from a talk given by Leon Lederman, maybe in the early 1970s.

quired civil construction, and (v) a stable political system. The first, and only, choice was Iran, then under the Shah.

The Muslim World The US Secretary of State, Hillary Clinton, announced[76] on 3 November, 2009 "New Initiatives to Bolster Science and Technology Collaboration With Muslim Communities Around the World" through a Science Envoy program that is part of President Obama's "New Beginning" initiative with Muslim communities around the world that he launched in a 4 June, 2009 speech in Cairo, Egypt. He pledged that the United States would "appoint new science envoys to collaborate on programs that develop new sources of energy, create green jobs, digitize records, clean water, and grow new crops." The envoys are Dr. Bruce Alberts,[77] Dr. Elias Zerhouni,[78] and Dr. Ahmed Zewail.[79]

It is not surprising that CERN has already begun such contacts through collaborations on the LHC experiments, including students and physicists from Oman, Saudi Arabia, Palestine, Jordan, Iran, and others including countries in Africa. High energy physics is an ideal platform for these contacts: it is universal, pure science without direct economic or military applications. In a sense, we owe a historic debt to the Islamic renaissance of the 8th–15th centuries AD[80] during which time much that Galileo and Newton discovered later had been understood to some degree.

North Korea A South Korean colleague and I once formulated a program to invite North Korean physicists to work at Korea University in Seoul, and at Iowa State University in Ames, and to go on to work at Fermilab on the D0 experiment. We had graduated to the point of writing a letter to the director of Fermilab on 11 September, 2001, after which we gave up. Would it have worked, in hindsight? It is not clear. It is still a good idea, not that it will work in the beginning, but eventually it will work, then everyone gains.

8.7 Problems

1. Maya Angelou, the acclaimed poet, once said "I've learned that people will forget what you said, people will forget what you did, but people will never forget how you made them feel." Translate this into rules for leading a collaboration.
2. Luis Alvarez (in his autobiography, *Alvarez*) once remarked that he had never fired anyone from his group, and that good people came to work in his group, and if they liked the work, they stayed. If the pressure was too much or the

76) http://www.state.gov/r/pa/prs/ps/2009/nov/131299.htm.
77) University of California, SF; former president of National Academy of Sciences.
78) Former director of National Institutes of Health.
79) Cal Tech; Nobel Prize Chemistry.
80) (See http://en.wikipedia.org/wiki/Islamic_Golden_Age.

work too difficult, they would go off to another group. Translate this into rules for managing a group.

3. Two individuals are fighting bitterly. It doesn't matter what the issue is, or who's right and who's wrong. They are both competent, even essential for the experiment. What do you do?

4. All organizations and political structures, whether scientific or not, have an organization chart defining the relationships and responsibilities between people and working units. It should be remarked that sometimes these definitions are not always clear, or not equally clear to everyone, but that is another issue. The three "org charts" in Figure 8.1 are meant to stimulate thinking, or at least to prompt thinking about how the structures of groups change over time. Where would you place the weekly meetings of the pre-ALEPH group, described by Steinberger?

5. Compare the similarities in the statements by Schopper, Delahaye, Wilson, Sessler/E.Wilson, and Kawamura. On what points do they differ?

6. Write a short proposal outline for scientific collaboration with North Korea suitable for presentation to the CERN Council.

Appendix A
Detectors and Instrumentation in 1960

One of my favorite books is the *Proceedings of the International Conference on* INSTRUMENTATION for HIGH-ENERGY PHYSICS held at the Lawrence Radiation Laboratory on 12–14 September 1960, 50 years ago, in about 300 pages. Thirteen current and future Nobel Prize winners in physics attended. But the interesting part is that we know which detectors and ideas worked and which didn't, and which were the germs for future instrumentation. The table of contents is reproduced here:

Session I	BEAM TRANSPORT AND RELATED TOPICS	Stanislas Winter
1.	High Magnetic Fields	Harold Furth
2.	Design and Construction of a System of Pulsed Magnets	Sydney D. Warshaw
3.	Some Specific Uses of High Magnetic Fields	Leona Marshall
4.	Cryogenic Magnet Coils for High-Energy Physics Experimentation	Richard F. Post
5.	Glass Cathodes in Vacuum-Insulated High-Voltage Systems	Joseph J. Murray
6.	Electromagnetic Mass Separation at Higher Energy	Myron L. Good

Particle Physics Experiments at High Energy Colliders. John Hauptman

ISBN: 978-3-527-40825-2

Session IIa	DETECTORS AND CIRCUITS, SPECIALIZED REPORTS	Herbert Anderson
1.	Low-Noise Fast Amplifier	Michiyuki Uenohara
2.	Tunnel Diode High-Speed Circuits	Hanoch Ur
3.	Nanosecond Counter Circuits	Robert M. Sugerman
4.	Performance of a Transistorized Nanosecond Counting System	Adrian C. Mellisinos
5.	Tunnel Diode Discriminator	Quentin A. Kerns
6.	Nanosecond Light Pulse for Coincidence Timing	Quentin A. Kerns
7.	A Fast 20-Channel Pulse-Height Analyzer Employing Line Coding	I.F. Quercia
8.	Pulse-Height Discriminator Employing Distributed Amplification	C. Infante
9.	Review and Evaluation of Fast Integral Discriminator Circuits	C. Infante
10.	A Coincidence-Anticoincidence Gas Cerenkov Counter	Dennis Keefe
11.	DISC: A Differential Isochronous Self-Collimating Cerenkov Counter	Arne Lundby
12.	Gas Cerenkov Counters for the K^+-Meson Channel of the Synchrophasotron	I.V. Chuvilo
13.	A Multichannel Focusing Cerenkov Counter	Robert A. Schluter
14.	Preliminary Evaluation of a Cerenkov Image-Amplifying Detector	Arthur Roberts
Session IIb	BUBBLE CHAMBERS, SPECIALIZED REPORTS	Luis W. Alvarez
1.	An Exploration of the Possibility of Employing Ultrasound Radiation to Sensitize a Bubble Chamber	Robert D. Sard
2.	A 5-Liter Rapid-Cycling Propane or Freon Bubble Chamber	Theodore Bowen
3.	Design of a 30-Liter Rapid-Cycling Hydrogen Bubble Chamber with Counter-Controlled Photography	Theodore Bowen
4.	A Pulsed-Resonant System Bubble Chamber Without Magnetic Field	Joe H. Mullins
5.	Xenon Bubble Chamber	John L. Brown
6.	Identification of Particles in Xenon Bubble Chamber Without Magnetic Field	I.V. Chuvilo
7.	Peformance of the Brookhaven National Laboratory 20-inch Hydrogen Bubble Chamber	Robert I. Louttit
8.	Reduction of Optical Distortion in Gas-Expansion Bubble Chambers	Harley C. Hitchcock
9.	Bubble Chamber Hodoscope	John A. Kadyk
10.	Use of Entrance Hodoscope for Particle Identification in Very-High-Energy Bubble Chamber Experiments	W. Selove
11.	The Principle of the Design of the CERN Propane Bubble Chamber	C.A. Ramm
12.	Cambridge Group Heavy-Liquid Bubble Chamber	Lawrence Rosenson
13.	The 1-meter Propane Bubble Chamber in a Magnetic Field	V.P. Dzhelepov
14.	A 200-Liter Heavy-Liquid Bubble Chamber	A. Rousset

Session III	PERFORMANCE AND CAPABILITIES OF BUBBLE CHAMBERS	Roger Hildebrand
1.	Relativistic Increase in Bubble Density in a $CBrF_3$ Bubble Chamber	B. Hahn
2.	Experience with a Large Hydrogen Bubble Chamber	Luis W. Alvarez
3.	Comparison Among Types of Bubble Chambers	Donald A. Glazer
4.	Application of the Helium Thermo-Cycle for a Liquid Hydrogen Bubble Chamber	V.M. Dobrov
5.	Gap-Length Measurement of Bubble Tracks	Charles Peyrou

Session IV	DEVELOPMENTS OF GENERAL INTEREST IN DETECTORS AND CIRCUITS	D.I. Blokhintsev
1.	A Discussion of Some Topics from Session IIa	Matthew Sands
2.	Cerenkov Counters	G. von Dardel
3.	A Velocity-Selective Gas Cerenkov Counter	G. von Dardel
4.	Spark Chambers	M.S. Kozodaev
5.	Applications of Solid-State Devices for High-Energy Particle Detection	Luke L. C. Yuan
6.	Handling of Counter Data	Clyde Wiegand
7.	Counter Data Recording System	W.A. Higinbotham
8.	The Use of a Sodium Iodide Luminescent Chamber to Study Elastic and Inelastic Scattering	Martin Perl
9.	Measurement of Particle Velocity with a Filamentary Chamber-Image Intensifier System	Kenneth Lande

Session V	TECHNIQUES INVOLVING RARE PROCESSES: EXPERIMENTAL TECHNIQUES AT ENERGIES ABOVE 20 BEV	Leland Hayworth
1.	Neutrino Experiments	
a.	Some Theoretical Implications of High-Energy Neutrino Experiments	C.N. Yang
b.	Remarks on High-Energy Neutrino Interactions	D.I. Blokhintsev
c.	A Neutrino Facility for the ZGS	Frederick A. Reines
d.	A Neutrino Detector for Use at the Brookhaven AGS	Leon Lederman
e.	The Possibility of the Detection of Neutrino Interactions in the CERN Heavy-Liquid Bubble Chamber	C.A. Ramm
f.	The Catholic University Neutrino-Detection System	Clyde L. Cowan
g.	High-Intensity Neutrino Beams	S. Courtenay Wright
h.	Possibility of High-Energy Neutrino Measurements with Cosmic Rays	Kenneth Greisen
2.	Features of the CERN Proton Synchrotron of Interest to Experimenters	M.G.N. Hine

Session VIa	REDUCTION OF DATA FROM BUBBLE CHAMBER FILM	L. Kowarski
Introduction		L. Kowarski
1.	Capabilities and Limitations of Present Data-Reduction Systems	Hugh Bradner
2.	Some Recent Developments in Data Reduction	A.M. Thorndike
3.	Recent Developments in Europe in Bubble Chamber Data Reduction	Y. Goldschmidt-Clermont
4.	Automatic Measuring Device for Bubble Chamber Photographs	S. Ya. Nikitin
5.	A Method for Faster Analysis of Bubble Chamber Photographs	Paul V.C. Hough
6.	The Spiral Reader Measuring Projector and Associated Filter Program	Bruce H. McCormick
7.	Automatic Scanning and Measuring of Bubble Chamber Negatives	Bruce H. McCormick
8.	Equipment for Fast Analysis	Jerome A. Russell
9.	Computer Programs and Uses	Arthur H. Rosenfeld
10.	Appendix: Spatial Reconstruction of Particle Tracks in Bubble Chambers	Frank T. Solmitz
Session VIb	DISCHARGE CHAMBERS, BEAMS, AND MISCELLANEOUS	T. Gerald Pickavance
1.	The Microwave Discharge Chamber – A New Type of Particle Detector	Shuji Fukui
2.	The Appearance of a Discharge in a Flat Controlled Counter Along a Particle Track	A.A. Tyapkin
3.	Studies of a Neon-Filled Spark Chamber	James W. Cronin
4.	Thin-Foil Discharge Chambers	Donald I. Meyer
5.	Properties of a Parallel-Plate Spark Chamber	Bruce Cork
6.	Face-View Pulsed Spark Chamber for Visual Location of Rapidly Successive Particle Tracks	Joachim Fischer
7.	Proposal for a Scintillation Coordinate Detector	Ernst Heer
8.	A High-Intensity μ-Meson Beam from the 600-Mev CERN Synchrocyclotron	Ansel Citron
9.	Some Features of Beam-Handling Equipment for the CERN Proton Synchrotron	C.A. Ramm
10.	A New 800-Mev/c K^- Beam of High Purity at the Bevatron	Peter Schlein
11.	Use of Generalized Amplitude and Phase Functions in Designing Beam-Transpsort Systems	K.G. Steffen
12.	Reliability of Beam Monitors	Franz A. Bumiller
13.	Precision Rotating-Coil Fluxmeter	Franz A. Bumiller
14.	A Two-Spectrometer System for High-Energy Electron-Scattering Studies	R. Hofstadter
15.	Remarks on the Use of a Solenoidal Iron-Free Spectrometer in High-Energy Electron Physics	R. Hofstadter
16.	Beam-Viewing Camera Using Rapid Development Film	Robert D. Sard

There was an obvious emphasis on bubble chambers in 1960, which were the newest and most glamorous particle detectors. Interestingly, bubble chambers are making a comeback in dark matter searches where pattern recognition and small spatial details are still important.

In addition, the wide range of technologies, techniques, and innovations is apparent in these talks. People were making new electronic circuits, pushing magnetic fields and their uses, bragging about their new chambers, advertising their beam lines, and speculating rather extensively on neutrino experiments.

Also apparent in these talks are the detectors that were forerunners in future Nobel experiments by Lederman/Schwarz, Cronin, Alvarez, and Hofstadter. Lederman and others were showing off their detectors and what they planned to do with them.

Problems then that are still being attacked today are the problems of "high magnetic fields" (where "high" is relative), "low-noise fast amplifiers" (where "low" and "fast" are relative), neutrino experiments (but not yet long-baseline), and the many incremental improvements in technique that make experiments better.

A number of Čerenkov detectors including (I guess) the first measurements of Čerenkov rings for imaging were shown here, in addition to particle identification and momentum measurement by Čerenkov light.

There have been instrumentation conferences before and many after this one, including the venerable Vienna series, the Elba conferences, and the newer TIPP (Technology In Particle Physics) series. It is my biased opinion that advances in instrumentation of all kinds are driven by experimental particle physicists, and by their demands on industry for better equipment, but this bias is less supportable today with rapid developments in silicon foundry works that exponentially generate better CPUs and memory chips. They now lead us in this area.

Appendix B
Detector Design Strategies

In truth, there is rather little in the previous eight chapters on any kind of recipe for designing a big detector, but it is clear that any design must begin with scientific goals and a very clear understanding of those goals. It is easy enough to say "I want good tracking" or "I want good calorimetry" or "I want both" and to specify criteria for each. It is more difficult to arrange detector systems that are not self-defeating in the sense that one detector will degrade the performance of another.

B.1
Comments about ATLAS and CMS

The LHC experiments CMS and ATLAS are very different in their outer muon measurement systems and in their electromagnetic calorimeters. The CMS experiment has emphasized electromagnetic calorimeter precision in its lead tungstate (PWO, $PbWO_4$) crystals with a $2.8\%/\sqrt{E}$ stochastic term compared to the ATLAS $10.1\%/\sqrt{E}$, in order to have the mass resolution on the Higgs mass in its decay

$$H^0 \to \gamma\gamma ,$$

to be able to separate this state from random $\gamma\gamma$ masses. This forced CMS into a noncompensating electromagnetic calorimeter with $e/h \sim 2.4$, with subsequent effects on the hadronic, or jet, energy resolution since the fluctuations in hadronic energy losses in the $e/h \sim 2.4$ part govern the overall jet energy resolution at $84\%/\sqrt{E}$ compared to ATLAS at $52\%/\sqrt{E}$. This is a clear physics choice and a bet on the existence of the Higgs. This physics choice is followed by a technology choice that is far more complicated, viz., (a) $PbWO_4$ crystals are more expensive, availability can be a problem, and "no two are alike"; (b) calibration and monitoring require $\Delta T \sim 0.1°C$ stability; and (c) radiation damage not only loses signal, but the losses are time-dependent. If the Higgs is discovered in this decay mode, CMS will be king; if not, then the continuing problems in calibration and radiation damage will extract a price.

For ATLAS, the electromagnetic calorimeter is a liquid Argon (LAr) accordion sampling calorimeter for which radiation damage is not a concern; however, (a) the slow signal is only partially collected in one crossing time (25 ns) and the signal itself is

Particle Physics Experiments at High Energy Colliders. John Hauptman

ISBN: 978-3-527-40825-2

not restored to baseline for 600 ns (24 bunch crossings), and (b) a temperature stability in the tens of mK range is required.

The muon systems are also dramatically different: CMS has a conventional Fe-chamber sandwich uniformly over nearly 4π steradians, while ATLAS has an iron-free detector with a toroidal field established by octets of superconducting loops, and consequently the field complexity will restrict muon acceptance to less than 4π. In both experiments the physics choice is the same (good muon physics) but the realizations are very different.

This shallow discussion of comparisons and problems with existing experiments illustrates that the specific designs and technology choices result in many levels of technical problems, all of which must be understood. For this understanding, nothing short of a full beam test is acceptable, as the excellent and extensive SPS test beams in the North Area at CERN testify.

This discussion also illustrates the problems in comparing details of different experiments with different physics goals. We don't really know what the correct physics goals are yet. Furthermore, the numbers given for resolutions are wrong; these are only the publicly stated resolutions for single pions in a test beam, but the real resolutions will depend on the material in front of these calorimeters and the confusion in events. It is even worse than this since in some cases the results of a test are only shown for a subset of the data, or the electronic noise is subtracted before the resolution is shown. And, what's more, single pions in beam tests are the easiest test case: QCD jets will be far more difficult. Nevertheless, all of these tests are essential.

B.2
Creativity vs. Conservatism

Almost always, new and better physics is achieved because someone had a better idea, either for a whole technology (e.g., Georges Charpak with multiwire proportional and drift chambers), or for an improvement or expansion in the implementation of a technology (e.g., Alvarez's big bubble chambers), or for a new *genre* of experiment (e.g., Ray Davis' swimming pool of cleaning fluid). Having a new idea is not so easy, and most new ideas are quickly forgotten as unworkable or even quite wrong. Once a colleague criticized a very creative physicist as "being wrong half the time." I thought, well, so he's right half the time, and that is phenomenally good.

Creativity is itself a fascinating subject, and I am not an expert. But I have noticed that very few people are creative, maybe because one has to be willing to be wrong and accept the (sometimes humiliating) consequences. I also think that being relatively new to an area, being an amateur, and in some sense being relatively ignorant helps with creativity since you are unaware of the limitations and boundaries in a problem. This in turn may explain why most truly intelligent physicists I have known appear less creative: maybe they veto their own bad ideas before expressing them.

The easy and safe way to design a detector is to use only those technologies that are familiar, maybe improve them a bit, but still they're known and safe. There is nothing wrong with this approach, and a functioning detector is almost guaranteed.

In the 4th concept there were so many creative people that at one point we actually discussed not presenting a new idea at the next meeting for fear that we would be thought a bunch of crazies. Some of our colleagues still thought so, however.

In one area, there were both creative and conservative people, and I encouraged them to work together. Maybe I was thinking of finding an "average" that was both creative and safe. This did not work. The personalities were fundamentally too different.

I firmly believe that in the early stages of brainstorming on a big experiment creativity should be encouraged in every respect, even when an idea initially appears to be stupid. It is also important that a social structure *not exist*, that there not be an organization chart in the beginning stages of an experiment. There will be plenty of time for organization charts later. One has only to read Steinberger's discussion of the pre-ALEPH group to guess that the weekly meetings had little or no structure.

During the prototyping and beam-testing middle stages – a very dynamic and fluid period where all the good ideas collide with the data – organization is necessary to coordinate the people and the institutional commitments that are required. The first introduction of "quality control" and funding levels comes near the end of these stages as a successful prototype is scaled up to a full detector.

The ending stages of construction and installation are the most critical – and are the stages about which I know almost nothing. I watched the TPC project at Berkeley flounder and almost fail before the director put an engineer in charge of every detail of the project, and he didn't care about your next new idea. He only cared about budgets and schedules, and the TPC was finished and did a decade of good physics, followed by the TPCs at DELPHI and ALEPH.

B.3
Main Design Issues for a Big Collider Experiment

As discussed in Chapters 6 and 7, the main global features of a big detector are the magnetic field, the tracking system, and the calorimeter system. The rest is detail. A vertex chamber is part of the tracking system; any add-ons to the calorimeter (preshower detectors, shower-max detector, barrel-end cap fillers, etc.) are part of the calorimeter system.

For a solenoidal magnetic field, either conventional as in CMS or dual-solenoid as in the 4th, it is my strong opinion that we should be making bigger solenoids and start now in their design and engineering. There are world experts, such as Alain Hervé *et al.*, who designed and built the CMS solenoid, which at its inception was a large and risky extrapolation beyond any previous solenoid.[81] A CMS physicist told me that initially the idea was to build the largest solenoid possible and then figure out how to put the tracking and calorimetry inside. This is a good strategy since many collider experiments have suffered from the mistake of putting material in front of the calorimeters, and even putting the solenoidal coil between the electromagnetic and hadronic calorimeters, with resulting poor energy resolutions in the calorimeters. The simplistic notion that the tracking system should have zero mass and the calorimeter system infinite mass is still a good one: no material before the particles hit the front face of the calorimeter, and no leakage after that.

This ideal requires radial space, both for the $1/\ell^2$ benefit to tracking momentum resolution and for the suppression of leakage and its fluctuations in the hadronic calorimeter.

Without larger solenoids, I am afraid we will be relegated to marginal calorimetry forever, unless the tracking is sacrificed for a 2-m-deep calorimeter, which is a bad idea. For a fixed radius solenoid, the tracking ℓ and the calorimeter depth are in direct, one-to-one competition, and a good overall physics detector will not be possible.

Because the $W^{\pm}$ and Z^0 are so central to all high-mass high-energy physics, and because they are so "democratic" in their decays to all partons of the standard model (all leptons, all quarks, and, therefore, essentially all composite particles in the hadronizing jets), it is obvious that any future detector must measure everything. The best calorimetry is well known and thoroughly published by the SPACAL, ZEUS, and DREAM collaborations, and the gross parameters of an excellent calorimeter are understood. Within calorimetry, and even within dual-readout calorimetry, there are many important and crucial issues not yet settled, although the DREAM collaboration[82] is systematically studying most manifestations of dual-readout calorimetry in fibers and crystals. It appears that a 2-m-deep calorimeter, equivalent to about $10\lambda_{\text{int}}$, is about right to provide the possibility of excellent calorimetry.

For tracking, magnetic fields are essentially limited to 5 T, 25% beyond the CMS existence proof, and the sagitta resolution depends on the combination of single-point resolutions and the number of such points. For the ILC there are three very

81) Alain Hervé, Domenico Campi, Benoît Curé, Pasquale Fabbricatore, Andrea Gaddi, Francois Kircher, and Stefano Sgobba, "Experience Gained From the Construction, Test and Operation of the Large 4-T CMS Coil", *IEEE Transactions on Applied Superconductivity*, Vol. **18**, No. 2, June 2008.

82) DREAM collaboration (2010): N. Akchurin[a], F. Bedeschi[b], A. Cardini[c], R. Carosi[b], G. Ciapetti[d], R. Ferrari[e], S. Franchino[f], M. Fraternali[f], G. Gaudio[e], J. Hauptman[g], M. Incagli[b], F. Lacava[d], L. La Rotonda[h], S. Lee[g], M. Livan[f], E. Meoni[h], A. Negri[f], D. Pinci[d], A. Policicchio[h], S. Popescu[a], F. Scuri[b], A. Sill[a], G. Susinno[h], W. Vandelli[i], T. Venturelli[h], C. Voena[d], I. Volobouev[a], and R. Wigmans[a],
[a] Texas Tech University, Lubbock (TX), USA;
[b] Dipartimento di Fisica, Università di Pisa and INFN Sezione di Pisa, Italy;
[c] Dipartimento di Fisica, Università di Cagliari and INFN Sezione di Cagliari, Italy;
[d] Dipartimento di Fisica, Università di Roma"La Sapienz" and INFN Sezione di Roma, Italy;
[e] INFN Sezione di Pavia, Italy;
[f] INFN Sezione di Pavia and Dipartimento di Fisica Nucleare e Teorica, Università di Pavia, Italy;
[g] Iowa State University, Ames (IA), USA;
[h] Dipartimento di Fisica, Università della Calabria and INFN Cosenza, Italy;
[i] CERN, Genève, Switzerland.

different tracking systems all of which (in simulation) achieved roughly the required momentum resolution of $\sigma_p/p^2 \approx 3 \times 10^{-5}$ (GeV/c)$^{-1}$:

SiD: all-silicon 5 layers with $\sigma_{\text{point}} \approx 5$ μm,
ILD: TPC with 250 points and $\sigma_{\text{point}} \approx 100$ μm, and
4th: cluster counting with 150 points and $\sigma_{\text{point}} \approx 50$ μm.

Of these, only the TPC has been extensively tested on particles, and none of the three has been subjected to the rates, track densities, and particle ionization loads expected at the ILC. Given that $B = 5$ T is maximal, that the realizable combinations of σ_{point} and N_{point} are limited roughly at those above for SiD, ILD, and 4th, there is only one parameter left: ℓ, the track arc length in the magnetic field.

Larger ℓ improves the momentum resolution by ℓ^2, but it pushes the calorimeter out farther and increases its volume as ℓ^2 (at constant depth) and increases the solenoid radius, which, for a constant aspect ratio, increases its stored energy as ℓ^3.

The only way to keep a detector on the real axis with respect to funding is to claim exceptional capabilities in precision tracking and precision calorimetry, all within a finite radial extent. In my opinion, these claims have not been demonstrated. However, there is still time before the start of the next machine construction, and many groups are working on these issues. With reference to Appendix A and recalling that “necessity is the mother of invention”, it is possible, but not at all guaranteed, that these ambitious capabilities will be attained.

My judgment is that physics is too precious to be limited and put at risk by the technology of large superconducting solenoids.

Therefore, it is my scientific opinion that the only way out is to begin the study, prototyping, and optimization of bigger solenoids. There is an additional benefit to R&D on solenoids: in every big collider detector, the superconducting solenoid is the single largest cost item, followed closely by the calorimeters.

It is easily conceivable that relaxing the radial restrictions on the calorimeter will lower its overall cost, so that R&D on larger solenoids will have a double return on investment. It is also true that other areas of science and technology can benefit directly from large-volume high-field solenoids, for example, large-volume NMR/MRI, versatile electrical energy storage, security, and so on. The 4th experts on big solenoids are Alexander Mikhailichenko (Cornell, LNS), Alexey Bragin (Budker Institute, Novosibirsk), Ryuji Yamada (Fermilab), and Masayoshi Wake (KEK), and among them are many good and original ideas.

The careful reader will notice that Steinberger's description of the early thinking and decisions on ALEPH followed this same path: first, decide on the magnetic field (large solenoid); second, decide on the tracking system (state-of-the-art TPC); third, decide on the calorimeters (using physics goals as the guide). In particular, the ALEPH solenoid was pushed to its then technical extreme, which immediately allowed a large-volume TPC[83] as the tracker.

B.4 Future Detectors

In the not-too-distant future, a laboratory director will form a program advisory committee (PAC) to consider the letters of intent and the formal proposals that have been submitted for experiments at the new collider. This machine will most likely be the ILC [1] if low-mass Higgs or supersymmetric particles are found at the LHC, or possibly CLIC [2] or the Muon Collider [3].[84] This PAC will be faced with multithousand-page proposals from groups that have been working for years on their designs.

It is my opinion that nothing short of a full tracking-calorimetry-muon system beam test (a so-called "slice" test) will be required in the proposal itself, even before consideration by a PAC. The community is getting ready for this eventuality with test beams at DESY, CERN, Fermilab, and an electron beam at SLAC. Such tests will not be easy since, in the case of PFA calorimeters, the performance of the fine-grained calorimeter is critically dependent on the performance of the tracking system, which requires a large-volume magnetic field.[85] Nevertheless, a "jet beam channel" can be mocked up to realistically test the sequence of vertex chamber-tracker-electromagnetic calorimeter-hadronic calorimeter-muon system that would be exposed to beam particles at all available energies.

There are some who disagree with the above paragraph and claim that a GEANT simulation that has been "validated" on beam test data is enough to guarantee the success of a big detector [60]. I think of this every time I am on shift at D0 watching the silicon strip tracking system of one million strips, with panels of green, orange, and red lights indicating whether a power supply is drawing less than a set current threshold, near the threshold, or above the threshold. If a channel trips repeatedly, we raise the threshold! This is slow radiation damage to the silicon, there are big variations throughout the silicon system, and none of it is in GEANT. Both the complexity and the surprises of a new detector are first appreciated in a beam test where, for the first time, beam particles spill across your device and a flood of pulses come down the coax cables. From these pulses alone you have to make sense of the detector.

83) The volume of the ALEPH TPC was 43 m^3, a big extrapolation beyond Nygren's original 6 m^3 TPC.

84) The thinking goes like this: the ILC is limited to $\sqrt{s} \approx 0.5$ TeV by the 35 MV/m RF gradient and the cost of the tunnel. If the LHC indicates a Higgs or supersymmetry at a mass above this energy, the ILC will not be built, but rather CLIC or a Muon Collider will be pursued with their potentially higher energy reach up to $\sqrt{s} \approx 3 - 5$ TeV.

85) In fact, the prototypes from R&D programs on large solenoids are perfectly suited for these beam tests.

B.5
Outline of Necessary Beam Tests with Particle "Jets"

For several years the DREAM collaboration has tested their calorimeters with "interaction jets" in addition to single pions [52]. A "jet" of particles spraying forward can be produced by a high energy charged pion interacting in a thin low-Z target in front of the calorimeter with thin scintillator plates in front and in back of the target. The trigger that an interaction took place producing a jet of particles is the coincidence of a mip signal in the upstream scintillator and a multiparticle signal (say, $>$ 10 mips) in the downstream scintillator. A 0.1-λ_{int} block of plastic (about 10 cm thick in the beam direction) allows an interaction efficiency of 10%, and the low-Z material does not absorb even the low energy particles.

These "jets" differ from QCD jets, of course, but they are completely sufficient as defined multiparticle jets on the one hand, and challenging enough in their complexity on the other, to test any calorimeter. These jets contain a broad momentum spectrum of most particles ($\pi^{\pm}$, $K^{\pm}$, K^0, $\bar{K}^0$, p, n, γ) and an angular spread not too different from quark and gluon jets of similar energies. We have used pion beams up to 300 GeV/c to produce these "interaction jets" without problems.

Low energy particles from the target can geometrically miss the calorimeter, but, for a dedicated test area, a solenoid modeled after the Project X target solenoid to capture pions from the target would catch most of these. A magnetic field could be tapered to be a "magnetic mirror" that would even redirect backward-going pions back into the calorimeter. In DREAM, we have learned a lot about the depth development differences between single pions and jets, and the differences in energy resolutions and leakage.

Beam tests of this character are more difficult, larger, and more challenging scientifically and technologically than many full physics experiments of only a few years ago, and much larger than any experiment discussed in Appendix A. These are ideal platforms for young physicists to work on instrumentation while gathering and analyzing physics data on the big running experiments at the LHC.

Appendix C
Glossary

A Few Useful Constants

c	$= 1/\sqrt{\mu_0\epsilon_0}$	$= 3 \times 10^8$ m/s $= 30$ cm/ns $= 1$ foot/ns	(velocity of light)
N_A		$= 6.02 \times 10^{23}$ per mole	(Avogadro's number)
m_e		$= 0.511$ MeV/c^2	(electron mass)
u	$= 1\text{g}/N_A$	$= 931.5$ MeV/c^2 $= \frac{1}{12} M(^{12}_{6}C)$	(unified atomic mass)
α	$= e^2/(4\pi\epsilon_0\hbar c)$	$= 1/137.036$	(fine structure constant)

Birks Suppression of Scintillation Light

A densely ionizing particle, for example a slow proton or α particle, will heavily ionize a column of material, and those *ionization* electrons with small kinetic energies can recombine with the positive ions in the column. This recombination suppresses the *scintillation* mechanism of an orderly deexcitation, and the optical light output is suppressed relative to the ionization *energy loss*.

Bremsstrahlung

Literally "braking radiation", the process through which energetic electrons that are accelerated in the high Coulomb field (Ze) of a nucleus emit photons. Therefore, bremsstrahlung is more pronounced in high-Z materials, and in fact a high energy electron will radiate most of its energy $(1 - 1/e)$ in 3.5 mm of W or 5.6 mm of Pb. All accelerated charges radiate, for example synchrotron radiation is a similar phenomenon, but it is due to acceleration in circular motion (v^2/r).

Calorimeter

A totally absorbing medium for the destructive measurement of particle energies. At colliders, they are generally divided into "electromagnetic" and "hadronic" calorimeters for the measurement of the two electromagnetic particles ($e^\pm$ and γ) and for the measurement of hadronic particles, predominantly $\pi^\pm$, $K^\pm$, K^0, p, and

Particle Physics Experiments at High Energy Colliders. John Hauptman

ISBN: 978-3-527-40825-2

n. The best description of these many possibilities is in [29]. There are also "weak interaction" calorimeters, in a sense, such as MINOS, T2K, NOVA, ICARUS, OPERA, and AMANDA, PIERRE AUGER, and other particle astrophysics experiments.

Čerenkov Radiation

Optical radiation in the deep blue and UV range that is generated by a charged particle traveling faster than the velocity of light in an optical medium, that is, in an optical medium of refractive index n; the velocity of light is c/n, and if a particle has velocity $v > c/n$, that charged particle will emit Čerenkov light. The generated spectrum is $dN/d\lambda \propto 1/\lambda^2$, and the light is instantaneous on the time scale of any experiment.

Compensation ("$e/h = 1$")

A hadronic *calorimeter* in which the average electromagnetic response is equal to the average hadronic (or nonelectromagnetic) response. In most calorimeters and in all crystals, e/h is larger than 1. The value of $e/h = 1$ is that fluctuations in π^0 relative to $\pi^\pm$ production in shower development do not matter for *energy resolution*.

Critical Energy (E_c)

The energy loss due to *bremsstrahlung* is proportional to E, but the *ionization energy loss* (at high energies) is proportional to $\ln E$, and the crossover energy where these two rates are equal is the "critical energy" E_c. Some useful relations are

$$E_c \approx \frac{dE}{dx} X_0 \quad \text{and} \quad E_c \approx \frac{610\,\text{MeV}}{Z + 1.24}(\text{solids/liquids}), \quad \text{or}$$
$$E_c \approx \frac{710\,\text{MeV}}{Z + 0.92}(\text{gases}).$$

All charged particles have a crossover in energy, and since it scales like the square of the mass, the critical energy for muons is

$$E_c^\mu \approx E_c^e \cdot (\frac{m_\mu}{m_e})^2 \approx 890\,\text{GeV}.$$

The critical energy governs the extinction and fundamentally determines the *energy resolution* of an electromagnetic shower as $1/\sqrt{N}$ for $N \approx E/E_c$.

Drift Chamber

An electric field, $\boldsymbol{E}$, drift region directs *ionization* electrons onto a sense wire (similar to *MWPC* sense wires), and from the known $\boldsymbol{E}$-field, electron trajectory, drift velocity of the gas, and the drift time, the spatial coordinate of the ionization is calculated. A drift chamber requires a "start" signal (which can be the beam crossing time), and the "stop" signal is provided by the avalanche at the sense wire. Typical spatial resolutions are 60–200 μm.

Dual Readout

Dual readout refers to hadronic *calorimeters* based on the simultaneous measurements of the electromagnetic and hadronic parts of a shower. Simply speaking, this means the electromagnetic $\pi^0 \to \gamma\gamma$ and the hadronic $\pi^{\pm}$ components. Since the instrumental response usually differs for these two components, independent measurements allow for a better *energy resolution*, a Gaussian response, and a calorimeter that is linear in hadronic energy. A concise description is by Wigmans (CERN-SPSC-2010-012; SPSC-M-771, "Dual-Readout Calorimetry for High-Quality Energy Measurements").

Energy Loss

All high energy particles traveling through atoms lose energy. For charged particles, it is predominantly *ionization* energy loss to the atomic electrons, formally given by the Bethe–Block expression, and numerically approximately equal to 1.5 MeV of energy loss per $\text{gm} \cdot \text{cm}^{-2}$ of material. Neutral particles such as n or K^0 (also νs) lose energy by scattering. Photons lose energy (all of it) by pair production. Energetic electrons can also lose energy by *bremsstrahlung*.

Energy Resolution (of a Calorimeter)

The response of a *calorimeter* is the distribution of the output, E, for the same repeated particle input. The root-mean-square (*rms*) of this distribution, σ_E, divided by the mean response, μ_E, is called the resolution, σ_E/μ_E. The rms is most legitimately determined by a fit of this distribution to a Gaussian

$$\text{Gaussian}(E; \sigma_E, \mu_E) = \frac{1}{\sqrt{2\pi}\sigma_E} e^{-\frac{1}{2}\left[\frac{E-\mu_E}{\sigma_E}\right]^2},$$

and the fit includes the tails of the distribution. If the tails are omitted from the fit, or only a portion of the distribution is fitted, the resolution is not Gaussian. Gaussian resolution is important: sums in quadrature require Gaussians, and every statistical theorem of any importance for physics either depends on or results in a Gaussian distribution.

Impact Parameter

There are three contexts in which the term "impact parameter" is used. In the calculation of ionization energy losses, the Coulomb force depends on the perpendicular distance from the passing particle's trajectory to an atomic electron, always referred to as the impact parameter. Similarly, in calculating the *rms* multiple scattering angle, the perpendicular distance from the passing particle to the atomic nucleus, the impact parameter, is needed to compute the Coulomb force.

Thirdly, in an experiment, either collider or fixed target, the particles that leave the interaction point should point back exactly to the interaction point within the measurement uncertainties on the track measurements. Tracks may not point back exactly to the vertex for two reasons: multiple Coulomb scattering in the target or beam pipe, and a track that comes from the decay of a massive long-lived particle produced in the interaction. The combination of center-of-mass momentum in the decay and the finite lifetime of the decaying particle result in a nonzero impact parameter of the track from the decay.

Interaction Length (λ_{int})

The nuclear interaction length is defined from the nuclear scattering cross-section for *inelastic* proton–nucleus collisions as

$$\lambda_{int} = \frac{A}{N_A \cdot \rho \cdot \sigma_{inel}},$$

that is, the mean attenuation distance due to nuclear collisions that will degrade the proton energy (since elastic and "diffractive" collisions result is nearly full-energy protons). The attenuation is described as $N = N_0 e^{-z/\lambda_{int}}$. The nuclear collision length is defined similarly but for the total proton–nucleus cross-section, σ_{total}, including the elastic cross-section, which is as large as 25% of the total. There are problems will all of this. First, the nuclear cross-sections are functions of energy with the $p\,p$ total cross-section varying by 25% from 1 to 10 GeV, and with the elastic part varying by a factor of 4. Second, the cross-sections for pion–nucleon scattering are about 2/3 of the proton cross-sections, so the interaction length is longer by 3/2, and the pion cross-sections vary even more with energy. For a hadronic *calorimeter*, the depth is usually measured in λ_{int}, and in practice at least $6\lambda_{int}$ are required for decent depth containment, but $10\lambda_{int}$ is better, relegating particle leakage to the irreducible leakage of νs and μs.

Ionization

Ionization refers to the removal of an electron from its atom, resulting in a positive "ion." Electrons are most easily removed by the transverse and transient Coulomb field of a passing charged particle that provides an impulse, Δp, to the electron.

Jet

The term "jet" is used to describe the collection of particles that materialize in the laboratory from a quark (q) or gluon (g) that is emitted in a parton level interaction. For example, the $W^{\pm}$ is a color singlet (has no color charge) and its hadronic decays are $W \to q\bar{q}$, leading to two "jets" in the detector. The sum of the four vectors of all the particles of both jets is approximately the four vectors of the $W^{\pm}$.

Landau Distribution

The Landau distribution is an analytic expression for the *ionization* energy losses of a particle in a material, and it is characterized by a high-side tail that represents the energetic ionization electrons that include the δ-rays that are evident in a bubble chamber photograph. The Landau distribution is not used much since discrete simulations of ionization are generally better for detector simulations.

LoI (Letter of Intent)

A letter of intent from a collaboration, or even from a protocollaboration, contains an author list, the scientific goals of an experiment, and a rather detailed description of the detector for evaluation by an advisory or review committee. It is more detailed than an "expression of interest" and less detailed than a formal proposal, which, in turn, is less detailed than a technical design report (*TDR*).

Linearity (of a Calorimeter)

The response of a *calorimeter* is the mean signal energy divided by the known beam energy, and the linearity, or nonlinearity, is the deviation of this response from a line that goes through the origin, that is, zero particle energy in must be zero signal out.

Mip (Minimum Ionizing Particle)

A *minimum ionizing particle* is any charged particle at a momentum corresponding to $\gamma\beta \approx 3$, where the *ionization* rate is a minimum, as shown in Figure 3.4. In principle, a *mip* does not exist since any particle at minimum ionization will quickly become more than minimum ionizing after a few μm of travel. In practice, a $\mu^{\pm}$ is often defined and used as a *mip* for calibration purposes.

Momentum Resolution (of a Tracking Chamber)

The same definitions as for *energy resolution* apply here, with the qualification that momentum, p, itself is not Gaussian distributed but rather inverse momentum, $1/p$ is Gaussian distributed, which is proportional to the sagitta of a track in a magnetic field. The sagitta is the primary measurement from which momentum is inferred from geometry.

Muon Chamber

Usually tracking chambers that are positioned within and outside the iron absorber that filters out the pions and other hadrons from the interaction. Particles that penetrate this iron shield, sometimes as deeply as 3 m, are predominantly muons.

MWPC (Multiwire Proportional Chamber)

A plane of typically 50 μm diameter "sense" wires at positive high voltage bounded on both sides by a zero-potential plane. The *ionization* electrons drift toward the positive sense wire with its increasing electric field, $E \propto 1/r$. When the kinetic energy gain of one electron, equal to the potential energy difference in one collision mean free path, exceeds the ionization potential of the gas, electron multiplication by collisional impact leads to an "avalanche" of electrons at the wire surface. One-half of the positive ions are created in the last mean free path, providing a pulse on the wire, which is the signal. For a wire spacing of $w = 2\,\text{mm}$, the rms spatial resolution is $w/\sqrt{12} \approx 0.6\,\text{mm}$.

Natural Units

There are several systems of natural units, one by Planck. In common usage, one usually means substitution of the fine structure constant α and the product $\hbar c$ in calculations. For example, any expression for electromagnetic energy or force involves a factor of e^2, for which one can substitute $\alpha = e^2/4\pi\epsilon_0\hbar c$. The product $\hbar c$ is often encountered, for example, in the de Broglie wavelength $\lambda = h/p = hc/pc = 2\pi\hbar c/pc$. In atomic physics, $\hbar c \approx 200$ eV-nm; in particle physics, $\hbar c \approx 0.2$ GeV-fm.

Occupancy

A detector with discrete channels will have some fraction of its channels activated by particles in one collision, in one beam crossing, or during one readout for that detector (which can be many beam crossings for a slow detector). It is therefore important to understand which of these is actually being calculated. A very "clean and quiet" detector will have an occupancy of 0.1%, or less. An occupancy of 50% would be very difficult to handle in a physics experiment since one-half of all channels will be "on" for every readout of an event. In tracking detectors, occupancy is usually driven up by background particles from either the beam or from some high-rate low-p_T process like $\gamma\gamma \rightarrow e^+e^-$ at an electron collider. One way to reduce occupancy is to make the denominator big, and this is one motivation for gigapixel vertex detectors, which are most at risk from beam-induced backgrounds.

PAC (Program Advisory Committee)

All laboratory directors appoint program advisory committees (PACs) to advise on the experimental program of the laboratory, and this committee can (and should) be very diverse in experimental expertise and broad in theoretical issues. In the end, a laboratory is often defined by its experiments.

PFA (Particle Flow Analysis)

Particle flow is the idea that a *jet* can be best reconstructed by using all the information from both the *tracking chambers* and the two *calorimeters* (electromagnetic and hadronic) together and, in particular, by using the better measured momenta of charged particles instead of their less well-measured calorimeter energies. The best description of the reconstruction methods is by Mark Thomson (arXiv:0907.3577v1 [physics.ins-det]), and descriptions of the test calorimeters are by the CALICE collaboration (arXiv:1004.4996v1 [physics.ins-det] and references therein). The so-called resolution of a PFA calorimeter is usually defined by its proponents in terms of a variable called "rms90", which is defined in arXiv:0907.3577v1 [physics.ins-det] and noted in this glossary under *rms*.

PMT (Photomultiplier Tube)

A remarkable instrument consisting of a photocathode for the conversion of a single optical photon into an electron (by the photoelectric effect) and the subsequent multiplication of the electron signal by repeated multiple electron emission by the kinetic energy impact on dynodes inside a vacuum tube. Many variations in all parameters (gain, size, speed, time resolution, geometry, etc.) are possible and widely used. PMTs are fast, low noise, high gain ($\sim 10^6$), and physically robust. The main weakness is their sensitivity to magnetic fields, which affect the electron trajectories inside the tube due to $\boldsymbol{F} = -e\boldsymbol{v} \times \boldsymbol{B}$, thereby reducing the gain and moving the tube away from its designed operating configuration.

Radiation Dose

The SI dose unit is a Gray (Gy), defined as the energy absorbed in a material (by *ionization* or excitation) in units of Joules per kilogram of mass, and 1 Gy = 100 rad. The equivalent dose is in units of Sieverts (Sv) and 1 Sv = 100 rem ("Roentgen equivalent man"). The radiation dose to detectors (and to humans) varies roughly as $\sim 1/r^3$ from the beam line, and varies by huge factors from machine to machine, where proton machines always have higher doses due to more nucleon–nucleon interactions ($p\,p$ cross-sections above 50 mb). The study of detector degradation due to radiation dose is very complex, and LHC studies are currently the best.

Radiation Length (X_0)

The radiation length is defined as the distance in a material over which the mean energy of an electron is degraded to $1/e$ of its initial energy. This mean energy loss is basically due to *bremsstrahlung* on the atomic nuclei of charge Ze, but including bremsstrahlung on the electrons and accounting for atomic screening effects in materials, a good relation for X_0 is

$$X_0(\mathrm{g/cm^2}) \approx \frac{716.4 \cdot A}{Z(Z+1)\ln(287/\sqrt{Z})}.$$

The radiation length always refers to electrons, and buried within the number 716.4 is the electron mass, but every charged particle radiates and the radiation length simply scales as $X_0 \propto m^2$ for any other particle. The radiation length governs the spatial scale for all electromagnetic phenomena in materials.

Rms (Root Mean Square)

The root mean square (rms) is a statistic with a precise definition. In an experiment with N randomly sampled values, x_i, $i = 1, \dots, N$, from an unknown distribution, the "mean" of the distribution is

$$\mu = \frac{1}{N}\sum_{i=1}^{N} x_i,$$

and the rms of the distribution, σ, is given by

$$\sigma = \sqrt{\frac{1}{N}\sum_{i=1}^{N}(x_i - \mu)^2},$$

that is, the square root of the mean squared deviation from the mean. For computational convenience, it is useful to calculate σ^2 as

$$\sigma^2 = \frac{1}{N}\sum_{i=1}^{N} x_i^2 - \left[\frac{1}{N}\sum_{i=1}^{N} x_i\right]^2 = \frac{1}{N}\sum_{i=1}^{N} x_i^2 - \mu^2 = \overline{x^2} - \overline{x}^2,$$

or the rms is "the square root of the mean of the square, minus the square of the mean",

$$\sigma = \sqrt{\overline{x^2} - \overline{x}^2}.$$

For a known continuous distribution, $f(x)$, the rms is given by

$$\sigma_x^2 = \frac{\int_{\mathrm{all}\ x} f(x) x^2 dx}{\int_{\mathrm{all}\ x} f(x) dx}.$$

In the *PFA* procedures, it is claimed that the energy response distributions are inherently non-Gaussian and that, therefore, a new statistic is required to describe the results. This statistic is called "rms90" and is defined as the "rms in the smallest region that contains 90% of the events." This means that 10% of the distribution is thrown out in such a way that the remaining 90% have a minimum rms. One should beware: this is not *resolution*.

Scintillation

All atoms are excited by passing high energy particles due to the transverse Coulomb force on the atomic electrons, the excited electrons being either ejected from the atom (*ionization*) or driven into higher atomic states (excitation). The deexcitation and emission of light is called "scintillation" when it is within an optical medium, copious, and fast (photoemission within a µs). A good description is given by the Particle Data Group (PDG) [49](Sections 28.3 and 28.4). When the atomic deexcitation process is slower, it can be referred to as phosphorescence or fluorescence.

Silicon Tracking Chambers

The depletion region in a semiconductor functions as an *ionization* collection volume, and when a charged particle ionizes the atoms of the medium, the ionization electrons and holes drift to their respective potentials of either conducting strips or pixels on the surface of the semiconductor. R&D in this area is very extensive and very important; see http://lpnhe-lc.in2p3.fr/ for the SiLC collaboration.

SiPM (Silicon Photomultiplier)

The SiPM ("silicon photomultiplier", but it has many names, for example, MPPC "Multi-Pixel Photon Counter", SPAD "Single Photon Avalanche Diode") is a recently invented device consisting of a matrix array of thousands of avalanche photodiodes (APD) acting as small Geiger cells, typically 10^3 within a few mm^2. A single optical photon will trigger one Geiger cell, delivering a large voltage signal. Two photons deliver twice the pulse height. Single-photon counting, high efficiency, and complete insensitivity to magnetic fields are the strengths of SiPMs, and dark rates (near the MHz range) and after-pulses (in the tens of a nanosecond range) are their weaknesses.

Tagging

Any distinguishing feature of an event, a *jet*, a track, or a collection of *calorimeter* towers can be "tagged." For examples, a track that has a nonzero *impact parameter* with the primary vertex can be "impact parameter tagged" as possibly coming from a B or D meson, a jet from within which a muon exits and penetrates the calorimeter can be "tagged" as possibly a b or c quark jet, and a narrow energy deposit in the calorimeter channels can be "tagged" as a possible $e^{\pm}$ or γ. The term tagging usually implies an incomplete identification.

TDR (Technical Design Report)

A very thorough and very detailed description of all the systems of a big detector. For the LHC detectors, the TDRs are several thousand pages in length, and cover all detector systems in great detail, including the trigger, data acquisition, and the software and computing resources. It is the best description of an experiment.

TPC (Time Projection Chamber)

A large-volume drift chamber in which the *ionization* electrons from a charged track are drifted in parallel electric $\boldsymbol{E}$ and magnetic $\boldsymbol{B}$ fields onto a detector plane. Thus, a TPC interrogates a three-dimensional volume of ionization with a two-dimensional detector. The first TPC was $6\,\mathrm{m}^3$ of Ar–CH_4 gas at nearly 10 atm. The $\boldsymbol{E}$ field drives the drift of ionization electrons in the gas with velocity $\boldsymbol{v} = \mu \boldsymbol{E}$ (μ is the electron mobility), and the $\boldsymbol{B}$ field suppresses transverse diffusion of the ionization electron clusters.

Tracking Chamber

Any collection of spatial measurement chambers through which multiple coordinates of a single particle passage can be associated to identify a charged track and specify its trajectory. The spatial measurement can be made by any device (from a Geiger counter to a silicon pixel chamber) that measures the spatial coordinate of a passing high energy charged particle. There are hundreds of possible tracking chambers: gaseous chambers include the drift chamber, *TPC*, *MWPC* (multiwire proportional chamber), and spark chambers; liquid chambers include bubble chambers; and solid chambers include silicon strips and pixels.

Vertex Chamber

A tracking chamber is usually positioned close to and around an interaction, or vertex point, to measure the trajectories of particles to high spatial precision. It is the main tracking device between the primary interaction and the main meter-sized tracking system. Its main physics purpose is to measure track impact parameters in order to tag weakly interacting states such as $D \to K\pi$ and $B \to D\pi$ with lifetimes and impact parameters of order $c\tau \sim$ 100–500 μm.

References

1 Barish, B. (2009), ILC/GDE report. International Linear Collider (ILC), http://www.linearcolider.org. The ILC is an e^+e^- collider based on superconducting RF cavities at 35 MV/m presently designed to reach $\sqrt{s} = 0.5$ TeV, with option to 1 TeV, in one interaction region. The "concept" detectors for the ILC are described in three Letters of Intent at http://www.linearcollider.org/physics-detectors/Detectors.

2 Delahaye, J.P. (2010) Compact Linear Collider (CLIC) (sometimes referred to as the CERN LInear Collider), http://clic-study.web.cern.ch. The CLIC design is a novel two-beam accelerator with RF cavities at 100 MV/m to reach $\sqrt{s} = 3$ TeV in one interaction region. The detectors (LCD) are described at http://lcd.web.cern.ch/LCD. See also, http://clic-study.web.cern.ch/clic-study/.

3 Geer, S. (2009), Muon colliders and neutrino factories. Also at: http://dx.doi.org/10.1146/annurev.nucl.010909.083736. The Muon Collider (http://www.fnal.gov/pub/muon_collider) design depends on an intense proton source, fast muon cooling, and superconducting RF (same as ILC) to reach $\sqrt{s} = 3$–6 TeV in a circular collider with two interaction regions. See also, https://mctf.fnal.gov/.

4 Mather, J.C. (2008) *The Very First Light*, Basic Books, 2nd edn. An excellent book in many respects: scientific, sociological and technological.

5 Halpern, P. (2009) *Collider: search for the world's smallest particles*, Wiley.

6 Sessler, A. and Wilson, E. (2007) *Engines of Discovery: A Century of Particle Accelerators*, World Scientific Publishing Co. Pte Ltd.

7 Feuersänger, C., (2009) *PGFPLOTS*, Bonn, http://sourceforge.net/projects/pfgplots/.

8 Carminati, F., Klapisch, R., Revol, J.P., Roche, C., Rubio, J.A., and Rubbia, C. CERN-AT-93-47. http://ppd.fnal.gov/experiments/e907/raja/energy_amplifier/convert.

9 http://lhc.web.cern.ch/lhc/.

10 Evans, L. (2009) *The Large Hadron Collider: a Marvel of Technology*, EPFL Press.

11 Lincoln, D. (2009) *The Quantum Frontier: The Large Hadron Collider*, Johns Hopkins University Press.

12 Panofsky, W.K. (1994) *Particles and Policy*, AIP Press.

13 Barish, B. (2009) ILC/GDE Report, TILC09, Tsukuba.

14 Delahaye, J.-P. (2009) CLIC09 Workshop. http://clic-study.web.cern.ch/CLIC-Study/.

15 SLAC Linear Collider (SLC) was the first prototype test of linear collider ideas and technologies; http://www2.slac.stanford.edu/vvc/experiments/slc.html.

16 See http://lhc.web.cern.ch/lhc/.

17 Grupen, C. and Shwartz, B. (2008) *Particle Detectors*, Cambridge University Press, 2nd edn.

18 Mokhov, N., Gudima, K., and Mashnik, S. *et al.* (2004) Physics models in the mars15 code for accelerator and space applications, *Technical Report*. See also, http://www-ap.fnal.gov/MARS/.

Particle Physics Experiments at High Energy Colliders. John Hauptman

ISBN: 978-3-527-40825-2

19 Sanami, T. *et al.* (2007) Radiation physics requirements in the IR, *Technical Report.*

20 Fermilab, Muon Accelerator Program (MAP), https://mctf.fnal.gov/mapproposal-r2f-mutac-1.pdf.

21 Perkins, D.H. (2000) *Introduction to high energy physics*, Cambridge University Press, 4th edn. This multi-edition textbook is still one of the best for a beginning student or non-expert.

22 Sjöstrand, T., Mrenna, S., and Skands, P. (2008) A brief introduction to PYTHIA 8.1. *Computer Phys. Commun.*, **178**, 852. Also available at arXiv:0710.3820.

23 Hansen, G. *et al.* (1975) Evidence for Jet Structure in Hadron Production by e^+e^--Annihilation. *Phys. Rev. Lett.* **35** 1609.

24 Brandelik, R. *et al.* (1979) Evidence for Planar Events in e^+e^--Annihilation at High Energies. *Phys. Lett.* **B86** 243.

25 Wilson, R. (1951) The range and straggling of high energy electrons. *Phys. Rev.*, **84**, 100–103.

26 Wilson, R. (1952) Monte carlo study of shower production. *Phys. Rev.*, **86**, 261–269.

27 Crawford, F., *et al.* (1957) Detection of Parity Nonconservation in Λ Decay *Phys. Rev.* **108**, 1102.

28 Orear, J., Rosenfeld, A., and Schluter, R. (1950) *Nuclear Physics*, University of Chicago Press, revised, tenth impression 1950 edn.

29 Wigmans, R. (2000) *Calorimetry – Energy Measurement in Particle Physics*, Oxford University Press. An excellent and completely thorough book on all aspects of calorimetry; essential reading for anyone seeking to understand or design and build a calorimeter.

30 Leo, W.R. (1994) *Techniques for Nuclear and Particle Physics Experiments: A How-To Approach*, Springer-Verlag, 2nd edn. An excellent readable book on all detector types, including electronics and techniques.

31 Ahmed, S.N. (2007) *Physics and Engineering of Radiation Detectors*, Academic Press. A very practical and thorough book.

32 Blum, W., Riegler, W., and Rolandi, L. (2008) *Particle Detection with Drift Chambers, 2ed*, Springer-Verlag, 2nd edn.

33 Bock, R. and Vasilescu, A. (1998) *The Particle Detector BriefBook*, Springer-Verlag. A nicely annotated glossary of terms.

34 Akchurin, N. *et al.* (2007) Contributions of cerenkov light to the signals from lead tungstate crystals. *Nucl. Instru. Meths.* **A582**, 474–483.

35 Molière, G. (1947) Theorie der streuung schneller geladener teilchen. i. einzelstreuung am abgeschirmten coulombfield. *Z. Naturforsch* **2a**, 133–145. A more accessible paper is by H.A. Bethe (1953) Molière's Theory of Multiple Scattering. *Phys. Rev.* **89**, 1256. Very recent updates verify the wide applicability of the Molière theory: R.N. Lee and A.I. Milstein (2009) Correction to Molière formula for multiple scattering. *J. Exp. Theor. Phys.* **108**, 977.

36 Bethe, H.A., "Molière's Theory of Multiple Scattering", *Phys. Rev.* **89** (1953) 1256.

37 Lee, R.N. and Milstein, A.I."Correction to Molière formula for multiple scattering" *J. Exp. Theor. Phys.* **108** (2009) 977.

38 CMS Collaboration (2010) Performance of the CMS Hadron Calorimeter with Cosmic Ray Muons and LHC Beam Data. *J. Inst.* **5**, T03012; arXiv: 0911.4991v3 [physics.ins-det].

39 Gluckstern, R. (1963) Uncertainties in track momentum and direction, due to multiple scattering and measurement errors. *Nucl. Instr. Meths.* **24**, 381.

40 Sauli, F. (1977) Principles of operation of multiwire proportional and drift chambers. *CERN 77-09.*

41 Nygren, D.R. (1976), Proposal for a PEP facility based on the time projection chamber.

42 Grancagnolo, F. (2007) The ultimate resolution drift chamber. *Nucl. Phys. Proc. Suppl.* **172**, 25.

43 Damerell, C., Willis, W., Vavra, J., Parker, S. *et al.* (2009) e-mail communications.

44 Savoy-Navarro, A., Silicon tracking for the Linear Collider (SiLC), http://lpnhe-lc.in2p3.fr/DOCS/.

45 Damerell, C. *et al.* http://physics.uoregon.edu/~lc/wwstudy/detrdrev.html.

46 Damerell, C. *et al.* (2008), Ilc vertex detector r & d. ILC-REPORT-2008-016.

47 Brau, J. (2009) Recent developments in detector technology. Also at arXiv:1003.2650v1 [physics.ins-det] 12 Mar 2010.

48 Wigmans, R. (2008) Energy measurement at the TeV scale. *New Journal of Physics* **10**, 025003. This is an excellent paper on resolutions, constant terms, compensation, segmentation, e/h, particle flow analysis, dual-readout, and the theoretical ultimate hadronic energy resolution achievable. This paper may be the best and most concise discussion of hadronic calorimetry. See also, *Calorimetry in the TeV Regime*, IEEE NPSS Lecture at the APS Workshop on the Future of High-Energy Physics, Snowmass, CO, 16 July 2001.

49 Amsler, C. *et al.* (2008) Review of particle properties. *Physics Letters* **B667**. The best and easiest source for most of particle physics, including accelerators, detectors, and the interactions of particles with matter: available at http://pdg.lbl.gov/.

50 Akchurin, N. (1998) On the differences between high-energy proton and pion showers and their signals in a non-compensating calorimeter. *Nucl. Instr. Meths.* **A408**, 380.

51 Acosta, D. *et al.* (1991) *Nucl. Instr. Meths.* **A308**, 481.

52 Akchurin, N. *et al.* (2005) Hadron and jet detection with a dual-readout calorimeter. *Nucl. Instr. Meths.* **A537**, 537–561.

53 Akchurin, N. *et al.* (2008) Dual-readout calorimetry with lead tungstate crystals. *Nucl. Instr. Meths.* **A584**, 304–318.

54 Akchurin, N. *et al.* (2008) Dual-readout calorimetry with crystal calorimeters. *Nucl. Instr. Meths.* **A598**, 710.

55 Akchurin, N. *et al.* (2009) Dual-readout calorimetry with a full-size BGO electromagnetic section. *Nucl. Instr. and Meth.* **A610**, 488–501.

56 Akchurin, N. *et al.* (2007) Measurement of the contribution of neutrons to hadron calorimeter signals. *Nucl. Instr. Meths.* **A581**, 643–650.

57 Wigmans, R. (1997) Quartz fibers and the prospects for hadron calorimetry at the 1% resolution level. Tucson CALOR

58 Morgunov, V. (2001) Energy-flow method for multi-jet effective mass reconstruction in the highly granular tesla calorimeter.

59 Brient, J.C. and Videau, H. (2002) The calorimetry at a future e^+e^- linear collider. ArXiv:hep-ex/02020041v1.

60 Thomson, M. (2009) Particle flow calorimetry and the PandoraPFA algorithm. *Nucl. Instr. Meths.* **A611**, 25–40.

61 The CALICE collaboration website: http://llr.in2p3.fr/activites/physique/ilc-calice.php/calice.html.

62 Abramowicz, H. *et al.* (1981) *Nucl. Instr. Meths.*, 180.

63 Milstene, C. and Pless, I. (July 1985), Improvement of the resolution of Pb/Cu calorimeter by software. APC Engineering Note 85-10, HERA.

64 Mazzacane, A. (2008) Jet reconstruction and physics performance with the 4th detector. *LCWS08 and ILC08*, 16–20 November 2008, Univ. Illinois, Chicago.

65 Uozumi, S. *et al.* (2002) *Nucl. Instr. Meths.* **A487**, 291. This calorimeter is also discussed in the GLD Detector Outline Document, http://physics.uoregon.edu/~lc/wwstudy/concepts/.

66 Behrens, U. *et al.* (1990) *Nucl. Instr. Meths.* **A289**, 115.

67 PD09 (2009) *International Workshop on New Photon Detectors*, 24–26 June 2009, Shinshu University, Matsumoto, Japan. http://www-conf.kek.jp/PD09/.

68 Frisch, H. (2008) Fast timing and tof in hep. Lyon Workshop on Picosecond Timing, Lyon, France, 17 October 2008.

69 Ramberg, E. (2010) Large Area Picosecond Level Photodetectors, Anti-proton Workshop, Fermilab, 22 May 2010.

70 Va'vra, J. *et al.* (2009) Beam test of a Time-of-Flight detector prototype, *Nucl. Instr. Meth.* **A606** 404.

71 Abramowicz, H. *et al.* (1981) *Nucl. Instr. Meths.*, **180**, 429.

72 CALICE (2009) Initial study of hadronic energy resolution in the analog HCAL

and the complete CALICE setup. *JINST, preprint.*

73 Della Negra, M.D. (2006) CMS physics: Technical design report. CERN/LHCC 2006-001, CMS TDR 8.1, 2 February 2006. Also, see D. Green, Ed., *The Leading Edge*, World Scientific, 2010.

74 The ATLAS detector at the LHC: http://atlas.ch/.

75 Drobychev, G. *et al.* (2009), Letter of intent from the 4th detector ("4th") collaboration at the international linear collider. Available at http://www.4thconcept.edu/4LoI.pdf.

76 Kraus, J. and ATLAS COLLABORATION (2010) First observation of electrons in the ATLAS detector. ATL-PHYS-PROC-2010-003, 15 January 2010, *XXth Hadron Collider Physics Symposium*, 16–20 November 2009, Evian, France.

77 Adams, I. *et al.* (2005) *Nucl. Instr. Meths.* **A538**, 281.

78 Edén, P., Gustafson, G., and Khoze, V. (1999) On particle multiplicities in three-jet events. *Eur. Phys. J C* **11**, 345–350.

79 Yang, C.N. (1960) Some theoretical implications of high -energy neutrino experiments. After dismissing three theoretical reasons for the existence of an intermediate-boson ("None of these reasons proves anything"), Yang says: 'To detect W it is important to realize that it has a very short lifetime ($< 10^{-17}$ sec) against decays into $\mu + \nu$, $e + \nu$, 2π, 3π, $K + \gamma$, $K + \pi$, etc. It is also important to realize that for theoretical reasons the W has a mass M_W larger than that of the K meson, and that it has spin 1. ... According to the schizon scheme there also exist two neutral W's: W^0 and $\overline{W}^0$. They cannot be produced by processes analogous to $[\nu + Z \to Z + \mu^- W^+]$. The only way to *experimentally* demonstrate their existence seems to be to produce W^0 and $\overline{W}^0$ in nucleon-nucleon or pion-nucleon collisions ... with a frequency 10^{-6} to 10^{-7} times that of the pions. These experiments will be rather difficult."

80 Adinolfi, M. *et al.* (2002) The kloe electromagnetic calorimeter. *Nucl. Instr. Meths.* **482**, 364–386.

81 DiBenedetto, V., Hauptman, J., Mazzacane, A., and Ignatov, F. (2010) Dual-readout, particle identification, and 4th. *Nucl. Instr. Meths. A*, doi: 10.1016/j.nima.2010.02.207.

82 Edwards, D.A. and Syphers, M.J. (1993) *An Introduction to the Physics of High Energy Accelerators*, Wiley-Interscience.

83 Reiser, M. (2008) *Theory and Design of Charged Particle Beams*, Wiley-VCH, Weinheim, Germany, 2nd edn.

84 Wille, K. (2000) *The Physics of Particle Acceleration*, Oxford University Press.

85 Persico, E., Ferrari, E., and Segre, S. (1968) *Principles of Particle Accelerators*, W.A. Benjamin, Inc.

86 Conte, M. and MacKay, W. (2008) *An Introduction to the Physics of Particle Acceleration*, World Scientific, 2nd edn.

87 Wilson, E. (2001) *An Introduction to Particle Acceleration*, Oxford University Press.

88 Chao, A. and Tigner, E.M. (1999–2009) *Handbook of Accelerator Physics and Engineering*, World Scientific, 3rd edn. Excellent; for experts.

89 Bruck, H. (1972) *Circular Particle Accelerators*, Los Alamos National Laboratory.

90 Green, D. (2010) *At The Leading Edge: The ATLAS and CMS LHC Experiments*, World Scientific. A collection of 16 papers by leading detector experts on CMS and ATLAS.

91 Budker, G. (1969) *Proceedings of the 7th International Conf. on High Energy Accelerators, Yerevan.* Extract in Physics Potential and Development of $\mu^+\mu^-$ Colliders: Second Workshop, (ed. D. Cline), AIP Conf. Proc. **352**, 4 (1996).

92 Skrinsky, A. (1971) *International Seminar on Prospects of High-Energy Physics, Morges.* Printed at CERN, unpublished; extract in Physics Potential and Development of $\mu^+\mu^-$ Colliders: Second Workshop, (ed. D. Cline), AIP Conf. Proc. **352**, 6 (1996).

93 Skrinsky, A. and Parkhomchuk, V. (1981) Cooling methods for charged particle beams. *Sov. J. Part. Nucl.* **12**, 223–247.

94 Steinberger, J. (2005) *Learning About Particles: 50 privileged years*, Springer.

95 Mönig, K. *et al.*. http://www.linearcollider.org/physics-detectors/Detectors.

96 Mönig, K. http://hepwww.rl.ac.uk/accel/forum/2006/moenig.pdf.

97 Berkelman, K. (2004) *A Personal History of CESR and CLEO: The Cornell Electron Storage Ring and Its Main Particle Detector Facility*, World Scientific Publishing Co. Pte. Ltd, Singapore.

98 Richter, B. (1976) *From the Psi to Charm – the Experiments of 1975 and 1976*, Nobel Lectures.

99 ILC forward region: http://www-zeuthen.desy.de/ILC/fcal/.

100 Hervé, A., Campi, D., Curé, B., Fabbricatore, P., Gaddi, A., Kircher, F., and Sgobba, S. (2008) Experience gained from the construction, test and operation of teh large 4-t cms coil. *IEEE Trans. on Applied Superconductivity*, **18**.

101 Alexander Mikhailichenko, private communication.

102 Wake, M., Yamada, R., and Tang, Z. (2009) Design study of iron free solenoid magnet for the 4th detector of ilc, *Technical Report*, Fermilab.

103 Gerwig, H. and Gaddi, A. CERN, private communication; also, see talk by Lucie Linssen, NIKHEF, 13 Nov 2009, http://www.cern.ch/lcd.

104 Ford, W. (1975) Iron ball, *Pep-186*. Result reported by M. Perl, "Review of Heavy Lepton Production in e^+e^- Annihilation", *1977 International Symposium on Lepton and Photon Interactions at High Energies*, Hamburg, Germany, August 25–31, 1977, SLAC-PUB-2022 (1977).

105 http://www.linearcollider.org/GDE/meetings/Snowmass-2005.

106 Hauptman, J. (2005), A fourth detector concept. At http://www.linearcollider.org/GDE/meetings/Snowmass-2005.

107 Akchurin, N. *et al.* (2004) Muon detection with a dual-readout calorimeter. *Nucl. Instr. Meths.*, **A533**, 305–321.

108 Akchurin, N. *et al.* (2005) Electron detection with a dual-readout calorimeter. *Nucl. Instr. Meths*, **A536**, 29–51.

109 Mikhailichenko, A. (2001) Do detectors need a yoke?, *Technical Report*, Cornell University, Laboratory of Nuclear Science (LNS) CBN 01-20.

110 Grancagnolo, F. and Miccoli, A. (2009) 4th loi: detector. TILC09, 17 April 2009, Tsukuba, Japan, http://tilc09.kek.jp/.

111 Bragin, A. (2009) Development of large superconducting solenoids. Budker Institure, 10 December 2009.

112 DiBenedetto, V. (2009) Implementing dual readout in ilcroot. 2009 Linear Collider Workshop of the Americas, Albuquerque.

113 This result by A. Cardini (INFN Cagliari) demonstrated that BGO would be easy.

114 Mazzacane, A. (2009) W/z separation. 2009 Linear Collider Workshop of the Americas.

115 DiBenedetto, V. (2008) Dual readout calorimetry in 4th concept.

116 Mikhailichenko, A. (2004) Why polarized positrons should be in the baseline of a linear collider, *Technical Report*, Cornell University, Laboratory of Nuclear Science (LNS).

117 Alexander, G. and Mikhailichenko, A. *et al.* (2008) Observation of polarized positrons from an undulator-based source. *Phys. Rev. Letts.* **100**, 210801. Based on an idea by Balakin and Mikhailichnko; also, see CLNS 06/1951.

118 Mikhailichenko, A. (2006) Few comments on the status of detectors for ilc, *Technical Report*, Cornell University, Laboratory of Nuclear Science (LNS).

119 Telnov, V. (2008) "Status of the Photon Collider", ILC ECFA Workshop, 9–12 June 2008, Warsaw; http://ecfa2008.fuw.edu.pl/.

120 Akchurin, N. *et al.* (2009) Neutron signals for dual-readout calorimetry. *Nucl. Instr. Meths*, **A598**, 422–431.

121 Ishii, M. *et al.* (2002), *Optical Materials* **19**.

122 Berners-Lee, T. and Fischetti, M. (1999) *Weaving the Web*, HarperCollins.

123 Larson, C. and LaFasto, F. (1989) *Teamwork: What must go right, what can go wrong*, Sage Publications, Inc.

124 Vardaman, S. and Stanford, J. (1994) *Statistical Methods for Physical Science*, Academic Press. Chapter 15, "Simulation of Physical Systems", J. Hauptman.

125 Akchurin, N. (1997) Beam test results from a fine-sampling calorimeter for

electron, photon, and hadron detection. *Nucl. Instr. Meths.*, **A399**, 202.

126 ILC-Report 2007.

127 TILC08, Joint ACFA Physics and Detector Workshop and GDE meeting on International Linear Collider 3–6 March 2008, Sendai, Japan, http://www.awa.tohoku.ac.jp/TILC08/.

128 Wigmans, R. (2006) Hadronic Simulation Workshop (HSS06), 6–8 September, Fermilab.http://conferences.fnal.gov/hss06/

Index

Particle Physics Experiments at High Energy Colliders. John Hauptman

ISBN: 978-3-527-40825-2